A PRIMER OF OFFSHORE OPERATIONS

Third Edition

By Ron Baker

Published by
PETROLEUM EXTENSION SERVICE
Division of Continuing Education
The University of Texas at Austin
Austin, Texas

in cooperation with
International Association of Drilling Contractors
Houston, Texas

1998

Catalog No. 1.10030
ISBN 0-88698-178-6

*The University of Texas at Austin is an equal opportunity institution.
No state tax funds were used to print or mail this publication.*

CONTENTS

FIGURES

34. On a bottle-type semisubmersible, partially flooded bottles submerge the rig below the water's surface but above the seafloor.
35. A semisubmersible at work
36. A column-stabilized semisubmersible rests below the waterline when in the drilling mode.
37. A drill bit has teeth to penetrate a formation to make hole.
38. Drilling mud is a special liquid essential to the drilling process.
39. Drilling mud jets out of the bit nozzles to lift cuttings off the bottom of the hole.
40. The mud pump picks up drilling mud from steel mud tanks and sends it down the kelly (or topdrive), drill pipe, drill collars, and bit. Mud and cuttings return to the surface up the annulus.
41. A heavy-duty mud pump
42. Mud and cuttings return to the surface via the annulus, the space between the pipe and sides of the hole.
43. A shale shaker screens drilled cuttings from the mud.
44. Hydrocyclones remove solids from mud.
45. A centrifuge spins mud at high speeds to remove solids.
46. A top drive (power swivel) is used to rotate the bit.
47. The kelly mates with the kelly drive bushing, which fits into the master bushing of the rotary table.
48. The kelly is made up on the swivel.
49. A horizontal well begins vertically and is deflected to horizontal.
50. A bent sub has a 1° to 3° bend in it to begin deflecting the hole from vertical.
51. A downhole motor laid out on a land rig's pipe rack
52. Several large diesel engines provide power for an offshore rig.
53. This generator, attached directly to an engine, makes electric power.
54. An electric motor provides power to a mud pump.
55. The driller controls rig power from a station on the rig floor.
56. Crown block, wire rope drilling line, traveling block, derrick, and drawworks are the basic hoisting system parts.
57. Drill collars are heavy pipe that put weight on the bit.
58. Racked in the derrick, this drill pipe is ready to be run back into the hole.
59. A typical drill stem assembly
60. A large hook, attached to the traveling block, supports the weight of the drill string.
61. Wire rope drilling line reeved through the traveling block's sheaves
62. The crown block at the top of a mast
63. The derrick on this drill ship supports the crown and traveling blocks and the drill stem.
64. Several lines reeved between the traveling block and the crown block increase hoisting capacity.
65. The drawworks is a large hoist.
66. Drilling line is wrapped around the spool (drum) of the drawworks.
67. A driller works at his station on the rig floor.
68. Standing on a small platform high in the derrick (the monkeyboard), a derrickhand guides the top of a stand of drill pipe. Note the safety harness to prevent a fall.
69. Three rotary helpers make up a joint of drill pipe.
70. A roustabout attaches a lifting sling to a crane's hoist line.
71. Usually, the roustabout foreman also operates the large cranes on the rig.
72. A ballast-control specialist stands at the ballast-control panel of a semisubmersible. The layout gives him a visual indication of the location and amount of ballast in the rig's pontoons.
73. Helicopter and boat companies transport personnel and supplies. *Left*, personnel transfer from the rig to the boat; *right*, a helicopter lands on the rig's helipad.
74. This station bill tells rig personnel how to recognize emergency alarms, and gives duty and station assignments.
75. Underway during a rig abandonment drill, personnel steer an escape capsule away from the rig.
76. An artificial gravel island and its drilling rig loom out of the fog in Alaska's Beaufort Sea. The sea is completely frozen and thus shows up as white behind the island.
77. Two joints of casing lie on a ramp, ready to be picked up and run into the hole.

78. A worker stands on the conductor casing that extends upward from the hole in the seafloor.
79. Cemented conductor casing lines the top part of the hole. More hole has been drilled out below the casing.
80. Several casing strings are run and cemented as the well reaches total depth. Note the names given to each string.
81. These blowout preventers are mounted on the casing.
82. Floating rigs are subjected to six motions caused by the wind, current, and wave action of the sea.
83. The temporary guide base is landed on the seafloor by drill pipe. Note the four guidelines running up to the rig.
84. A hole opener enlarges the hole made by the regular bit.
85. The guide frame centers the bit and hole opener into the opening in the temporary guide base.
86. A Squnch Joint™, a threadless connector used to make up large-diameter joints of casing
87. The permanent guide structure attached to the top of the foundation-pile housing is lowered on the guidelines. The foundation-pile casing enters the drilled hole, and the permanent guide structure seats on the temporary guide base.
88. The subsea BOP stack is lowered onto the permanent guide structure.
89. This marine riser system is connected to the top of the subsea BOP stack.
90. A riser tensioning system provides upward tension on the riser system.
91. These three piston-and-cylinder assemblies are part of the riser tensioning system.
92. Buoyant devices attached to riser joints augment the riser tensioner system.
93. This heave compensator assembly is attached to the traveling block. Note the two pistons and cylinders that offset heave motion.
94. The heave compensator keeps weight on the bit at a constant value in spite of the floater's up-and-down movement.
95. The curves on this well log can reveal information about the formation penetrated by the hole.
96. When made up on a core barrel, the core bit *(left)* cuts a long cylindrical sample of the formation, which is called a core *(right)*. Only a short portion of the core is shown.
97. Several cores are laid out for analysis at an onshore laboratory.
98. Special burners flare gas harmlessly into the atmosphere on this unit in the North Sea.
99. A platform tender is anchored next to a relatively small platform.
100. A self-contained platform houses all the drilling and production equipment, crew quarters, galley, offices, and recreation rooms.
101. A platform jacket is so tall, it is usually built and transported on its side.
102. A large crane hoists a deck section onto the jacket.
103. Five boats tow a concrete platform to a site in the North Sea.
104. Concrete cylinders arranged around the base of a concrete gravity platform provide storage space for oil.
105. A caisson-type platform rests in the ice-free waters of early summer in the Cook Inlet of Alaska.
106. The relatively lightweight jacket of a guyed-tower platform is supported by several guywires and clump weights.
107. A buoyant tension-leg platform is held on station by several tensioned tubes attached to the seafloor.
108. A jackup drills development wells on a small jacket *(right)*.
109. Two relatively small platforms replace a single large platform.
110. A subsea template allows several wells to be drilled, completed, and produced from a relatively small area on the seafloor.
111. Several directionally drilled wells tap an offshore reservoir.
112. A perforating gun is lowered into the wellbore and fired. Shaped charges perforate the casing, cement, and formation.
113. When a packer is set, it forms a seal between the outside of the tubing and the inside of the casing, causing reservoir fluids to flow into the tubing.
114. In a surface completion, equipment that controls the flow of hydrocarbons from the well is placed on a deck of the platform.
115. A wet subsea Christmas tree installed on the seafloor

116. This dry subsea Christmas tree is isolated from the water by a protective dome.
117. In a dissolved-gas drive reservoir, gas comes out of solution from the oil and drives oil to the surface.
118. In a gas-cap drive, the gas cap expands to lift oil to the surface.
119. In a water drive, water moves oil to the well and up to the surface.
120. In a combination drive, expanding gas and water lift oil to the surface.
121. This production platform retains the derrick used in drilling the wells.
122. In this subsea production system, oil flows from wells on the template, up the production riser, and to a special mooring buoy, where a tanker loads the oil.
123. A free-water knockout removes free water from produced fluids.
124. A photomicrograph of a water-in-oil emulsion reveals the small droplets of water that are dispersed through the oil.
125. A horizontal treater is used to remove most of the emulsified water from oil.
126. A glycol dehydration system removes most of the water from natural gas.
127. Two horizontal separators installed on this platform separate oil and gas.
128. Gas lift involves the injection of gas into a well to lower pressure at the bottom of the well. (A) Since the well is not producing, no liquids are flowing; (B) injected gas forces liquid out through tubing; (C) gas enters top valve and lightens liquid in tubing, causing liquid level to fall further; (D) well is producing fluids from the reservoir.
129. When water is used for pressure maintenance, it is injected into the reservoir to drive oil to producing wells. Some of the oil will be left behind, however.
130. Injecting a surfactant can recover additional amounts of oil from a reservoir.
131. Alternating injections of CO_2 and water can recover additional oil.
132. A mobile jackup rig services a well.
133. A wireline unit is used to make well repairs.
134. Gas-lift valves installed in side-pocket mandrels can be run and pulled with wireline.
135. A tie-in pipeline joins a trunk line running to a shore facility.
136. This lay barge is at work in the North Sea.
137. Protruding from the stern of this semisubmersible lay barge is the stinger, which supports the pipe as it enters the water.
138. Pipeline is wound onto the reel of this reel ship.
139. By using high-pressure jets of water, a bury barge digs a trench for a pipeline.
140. A single-point buoy mooring system loads oil into a tanker.

PREFACE

This book is for people who don't know very much about offshore oil and gas operations and who want to know more. We wrote it in simple language and, even though we used technical terms from time to time, we defined them in the text and in the glossary. Further, the text does more than simply describe an operation or method; it also tells why such an operation or method is necessary.

Even though the offshore oil industry came of age in the United States, this book is not confined to this country. Since operations are ongoing in the North Sea, Southeast Asia, Africa, South America, China, and the Arctic (to name a few places), the text covers techniques and equipment utilized the world over. In cases where the text gives dimensions, it gives both English and SI metric measurements.

PETEX gratefully acknowledges the many people without whose assistance and encouragement this book would not have come about. Tom Thomas of Sedco Forex Schlumberger rounded up many of the photos from various sources within his company. He also provided a great deal of technical input, as well as enthusiasm for the project. Dan Simon, Ed DuBroc, and Cohen Guidry of Nabors Drilling U.S.A., and Sundowner Offshore Services, Inc. (a Nabors Industries Company) went out of their way to get me on both land and offshore rigs to shoot photos and learn more about the complex world of offshore operations. Thanks too, to Lee Hunt, executive vice-president of the International Association of Drilling Contractors (IADC), for his organization's cooperation. Ken Fischer of IADC persuaded a couple of their members to read the old edition to give us help in updating the new one. John Alterman, Reading & Bates Drilling Company, and A.J. Guiteau, Diamond Offshore, are two who read the old text and made numerous, helpful suggestions. While we cannot mention everyone who helped (there were just too many), we especially thank the many oil companies, drilling contractors, well service and workover contractors, and service and supply companies who allowed PETEX to visit and photograph a large number of their offshore installations, gave us permission to use their material, and who read and critiqued the text. Without the assistance of the industry, this book would not have been possible.

Members of the PETEX staff put this book together, which is one tough job. Kathryn Roberts edited and organized photos, drawings, and material from many sources. Debbie Caples laid out and typeset the book in her usual superlative manner. Doris Dickey read, reread, and reread again the manuscript to track down typos and other errors.

In spite of the assistance PETEX got in writing and illustrating this primer, PETEX is solely responsible for its contents. Keep in mind, too, that while we made every effort to ensure its accuracy, this manual is intended only as a training aid, and nothing in it should be considered approval or disapproval of any specific product or practice.

Ron Baker
Director

INTRODUCTION

People use oil and gas more than any other source of energy. From oil, refineries make or extract gasoline, diesel fuel, and lubricants. Petrochemical plants make plastics and fertilizers. Natural gas heats our homes and fires steam generators to make electricity. Without oil and gas, everyone's life would be very different.

The petroleum industry produces oil and gas from special layers of rocks called reservoirs. Like a multilayered cake, additional beds of rock lie above and below these reservoirs. And, like the frosting on a cake, a relatively thin layer of ground sometimes covers the rock layers. On the other hand, the "frosting" may not be dry land; it may be water instead. Since oceans and seas cover about three-fourths of the earth, it is no surprise that water also covers rock layers.

Operating in oceans or seas—offshore—presents special problems to oil producers that they do not have to face on land sites. This book examines many of the special conditions the marine environment imposes on finding, producing, and transporting oil and gas.

First Offshore Operations in the U.S.

In the United States, offshore oil and gas operations began in the late 19th century. Edwin Drake drilled the first oilwell in the U.S. in 1859. He did it on a piece of land near Titusville, Pennsylvania. It was only thirty-eight years later, in 1897, that another enthusiast drilled the first offshore well in the U.S. He drilled it off the coast of Southern California, immediately south of Santa Barbara.

In the late 1800s, a group of people founded the town of Summerland, California. The founders picked the site because of its pleasant, sunny climate. Coincidentally, it also had numerous springs. These springs did not, however, produce water: crude oil and natural gas bubbled out of them.

Since Summerland could use gas to light its homes and businesses, and since oil could provide income, the city's residents took an interest in efficiently producing the springs. One citizen, H.L. Williams, was knowledgeable about extracting oil from the earth. So, just as Drake had done earlier in Pennsylvania, Williams drilled wells in the vicinity of the springs. The wells allowed him to extract more oil than if he had simply dammed up the springs. These early wells were successful and, as a result, he and others drilled many more in the area.

After drilling a large number of wells, these early oilmen noticed that those nearest the ocean were the best producers. Eventually, they drilled several wells on the beach itself. But, at this point, the Pacific Ocean stymied them. Experience convinced them, however, that more oil lay in the rock formations below the ocean. The question was how to drill for it.

Williams came up with the idea of building a wharf or a pier and erecting the drilling rig on it. The idea worked. His first offshore well, drilled from a wharf made of wooden pilings and timbers, extended about 300 feet (90 metres) into the ocean. On the end of the wharf, Williams erected a drilling rig and used it to drill the first offshore well in the United States. As expected, it was a good producer and before long the entrepreneurs built several more wharves (fig. 1). The longest stretched over 1,200 feet (nearly 400 metres) into the Pacific.

Figure 1. The first offshore wells in the United States

The Scope of Offshore Operations

Today, offshore activities take place in the waters of more than half the nations on earth. And no longer do primitive, shore-bound wooden wharves confine offshore operators. Instead, they drill wells from modern steel or concrete structures. These structures are, in many cases, movable. What is more, they can float while being moved, and often while drilling. Further, offshore rigs have drilled in waters over 7,500 feet (over 2,200 metres) deep and as far as 200 miles (over 300 kilometres) from shore. Offshore drilling and production have progressed far beyond those early efforts at Summerland.

Offshore work today involves a wide range of technologies. These technologies are similar in many cases to those used to find, produce, and transport oil and gas on land. Offshore activities include, however, additional technologies that relate to a marine environment. Unlike oil operations on land, offshore operations involve meteorology, naval architecture, mooring and anchoring techniques, and buoyancy, stability, and trim.

Drilling and producing oil and gas wells are important phases of offshore operations, but the scope goes further. Offshore operations also include exploring—looking for likely places where oil and gas may exist in the rock formations that lie beneath the surface of the oceans, seas, gulfs, and bays. In addition, offshore operations include transporting oil and gas—moving them from their points of production offshore to refineries and plants on land.

1
OIL AND GAS

Oil people often call crude oil petroleum. *Petroleum* comes from the Latin word for rock, *petra*, and the old Greek word for oil, *oleum*; petroleum literally means "rock oil." Petroleum is a good name for crude oil because it really does come from rock. The rocks in which oil companies find petroleum have special characteristics.

Explorationists—those who look for oil and gas—sometimes have a hard time finding the special rocks. Usually, many other rock layers cover and bury them below the earth's surface. Like petroleum, natural gas also comes from special rocks buried far below the earth's surface. Frequently, gas occurs with oil.

Characteristics of Oil and Gas

Oil and gas are *hydrocarbons*. That is, oil and gas contain only two elements: hydrogen and carbon. (An element is a substance that consists of atoms of only one kind. At present, about 112 elements exist. Almost 100 of them occur naturally; nuclear scientists make the rest in laboratories.) Other substances, such as sulfur, carbon dioxide, nitrogen, and salt, may also exist with gas and oil, but the gas and oil are hydrocarbons.

CHEMICAL MAKEUP

Even though only hydrogen and carbon make up hydrocarbons, a hydrocarbon's chemical structure is not necessarily simple. Hydrogen and carbon have a great attraction for each other and can arrange themselves in simple or very complex ways.

Natural gas tends to be less complex than crude oil. As it comes out of a well, it is mainly methane (CH_4), which is the simplest hydrocarbon. Natural gas frequently contains heavier hydrocarbons such as ethane (C_2H_6), propane (C_3H_8), and butane (C_4H_{10}). In addition to hydrocarbon compounds, natural gas may contain other gases, such as nitrogen, carbon dioxide, helium, hydrogen sulfide, and water vapor.

Crude oils can be complex. They often contain not only simple hydrocarbons, such as methane, but also complicated liquid and sometimes solid hydrocarbons. The complex structure of oil can keep even advanced chemists occupied with studying it.

PROPERTIES

Methane (natural gas) is odorless, colorless, less dense than air, and flammable. Because natural gas is naturally odorless and so flammable, gas companies add a chemical to make it smell bad before selling it. This odorant allows you to smell leaking gas and thus avoid accidents.

Crude oil varies widely in appearance. Its color can range from pitch black to pale straw. In weight, or density, it ranges from very dense—denser than water—to very light, perhaps only three-fourths as dense as water. Its viscosity, or resistance to flow, ranges from solid, which does not flow at all, to very thin liquid—almost like water. Crude oil's odor can range from very pungent to almost odorless. And, of course, it is flammable, which is partly why it is so valuable.

Characteristics of Rocks

Many people have the mistaken idea that oil occurs in underground lakes and rivers. In fact, both oil and gas occur in tiny openings in rocks. These rocks are usually buried deep beneath the surface. Rocks that hold oil and gas actually have small voids, or openings, in them, even though they may look solid to the naked eye. Geologists call these openings *pores*. If oil and gas exist in a rock, they will be in the rock's pores.

Some rocks not only have pores but also small passageways that connect the pores. A rock that has connected pores is *permeable*. Permeability in a rock allows any hydrocarbons in the rock's pores to move from one pore to another (fig. 2).

Assume that a rock has pores and is permeable. Also assume that oil or gas occurs in the pores. In such cases, those who seek oil and gas have a promising situation: a well drilled into the rock may produce oil or gas. To make it worth the expense of drilling the well and extracting the hydrocarbons, however, the rock holding the hydrocarbons must cover a large area. Also the large area must hold enough hydrocarbons for a company to recover its costs and make a profit.

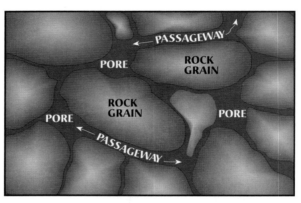

Figure 2. A permeable rock allows hydrocarbons to flow from pore to pore within the rock.

Types of Rocks

Geologists—those who study the earth and its composition—classify rocks into three types: igneous, metamorphic, and sedimentary. Molten material from the interior of the earth that solidified as it cooled in the earth's crust or on the surface formed *igneous* rocks. Granite is an igneous rock.

When heat, pressure, or chemical action in the earth changes an existing rock it becomes a *metamorphic* rock. Marble, a good example of a metamorphic rock, comes from the sedimentary rock limestone.

Most often, oil and gas appear in sedimentary rocks. Preexisting materials—sediments—form *sedimentary* rocks. Wind, water, or ice carried the sediments and deposited them in another place. Geologic forces then changed the sediments into rock.

As an example, think of a river flowing over igneous rocks. Slowly and surely the water erodes the rock over which it flows. The river picks up the eroded material, or sediment, and carries it to an ocean.

In the ocean, the sediment falls to bottom and, given enough time, builds to form a very thick layer. Eventually, those layers deposited first become buried very deeply below the seafloor. The high heat and the high pressure of the great depths transform the sediments into sedimentary rock.

Shale, sandstone, and limestone are common types of sedimentary rock. Fine sediments such as mud or clay make up shale. Coarse-grained sediments such as sand make up sandstone. Calcium carbonate, which mainly comes from the skeletal remains of marine life, makes up limestone. Geologists often refer to limestone as *carbonate* rock; they refer to shale and sandstone as *clastic* rocks. Pieces of preexisting rocks make up clastic rocks.

Of all the sedimentary rocks, petroleum geologists are the most interested in sandstone and limestone because they find most of the oil and gas in such rocks. Shale is also important, however. Since it is virtually impermeable, it can prevent the escape of hydrocarbons from sandstone or limestone.

Origin of Oil and Gas

No one can say for sure exactly how oil and gas formed, but scientists have worked out several theories using available evidence. The *organic theory* is the most popular. It states that oil and gas come from the remains of plants and animals that lived and died in the seas millions of years ago. The organic theory takes its name from the fact that plants and animals are organisms.

Most scientists think that the plants and animals that gave rise to oil and gas were microscopic. This is because it took millions of tiny plants and animals to create hydrocarbons and extremely large numbers of tiny plants and animals lived in the ancient seas.

Rivers carried plants and animals to the sea where they died and settled to the bottom. The rivers also carried silts and muds to the sea. The sea itself was home to countless microscopic plants and animals.

On the bottom these small sea organisms mixed with the silt, sand, and other dead river life to form a rich organic mixture. The silts and sands prevented any oxygen dissolved in the water from affecting the dead organisms. Without oxygen, the organic matter could not decay normally.

Thus, over an extremely long period, a thick blanket of undecayed organic sediment collected on the seafloor. As more time passed, large amounts of additional sediments covered the organic material. The great weight of these overlying sediments put tremendous heat and pressure on the organic matter buried below.

The heat and pressure transformed the organic matter into oil and gas. What is more, the heat and pressure transformed the surrounding inorganic material into sedimentary rock.

Migration and Accumulation of Oil and Gas

The weight of overlying rocks created great pressures on the newly formed oil and gas. These pressures forced them to move through any surrounding permeable rock. Fluids move from an area of high pressure to an area of low pressure. So, as the petroleum moved through the permeable rocks, it tended to move upward, seeking the surface. Surface pressure is much lower than subsurface pressure.

In some cases, petroleum actually reached the surface, like the oil springs that H. L. Williams drilled next to in California. But in most cases, petroleum traveled upward until an *impervious*, or impermeable, rock barrier blocked it below the surface. Once blocked, it accumulated in the rocks below the impervious barrier.

Meanwhile, on the surface was land or water. Sometimes, the surface environment was tropical; in other instances it was arctic. Similarly, the surface overlying the hydrocarbons could be near a shore or quite distant from it. Further, the surface could be accessible or inaccessible. Regardless of what the surface was like, the hydrocarbons lay trapped in rocks, usually thousands of feet (metres) down, awaiting discovery.

Traps

A *trap* is an arrangement of rock layers that contains hydrocarbons. Essentially, a trap has an impervious rock layer that prevents the escape of hydrocarbons from a porous and permeable layer. Put another way, hydrocarbons occur in the porous layer, and the impervious layer keeps them there. Often, the impervious layer is shale, while the porous layer is sandstone or limestone.

Strong forces in the earth have created many types of traps. Most are the result of a geological event that deformed the original rock layers. For example, earth forces can fold layers of rock that were originally deposited horizontally into arches or troughs. Earthquakes can move large blocks of earth upward, downward, or sideways. Over very long periods of time, erosion from wind, rain, and ice can wear down even the mightiest of mountains. Molten rock can thrust upward and bend and break thousands of feet (metres) of overlying rock beds.

All these geological forces and more can radically alter not only the earth's surface but the subsurface as well. These subsurface alterations can serve as potential traps for hydrocarbons.

The basic types of traps are those caused by folding, faulting, unconformities, domes or plugs, changes of permeability within a formation, and combinations of these.

FOLDING

Folding creates *anticlinal traps* (fig. 3). Earth movements folded the rock layers upward into an arch. Hydrocarbons then migrated into one of the anticline's porous and permeable layers and accumulated in its upper portions. An overlying impervious bed sealed the hydrocarbons in the porous layer.

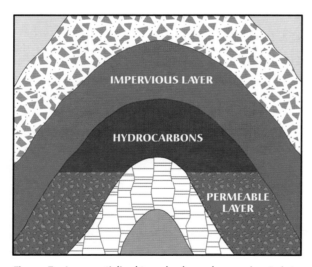

Figure 3. In an anticlinal trap, hydrocarbons migrate into a porous and permeable layer and are trapped there by an overlying impervious layer.

FAULTING

A *fault* is a fracture, or break, in rock layers that movements in the earth create. In fault traps, the reservoir—the rock layer that holds the hydrocarbons—is usually on one side of the fault. On the other side, an impervious layer moved opposite the reservoir, which prevented further migration of the hydrocarbons (fig. 4).

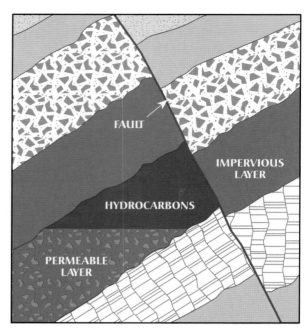

Figure 4. In a fault trap, an impervious layer moves opposite a porous and permeable layer, thereby trapping hydrocarbons.

UNCONFORMITY

An *unconformity* is a lack of continuity between rock beds, or strata. Sometimes, the unconformity can serve as a trap for oil or gas. An impermeable layer on top of the weathered surface of the lower beds prevents the upward escape of hydrocarbons (fig. 5).

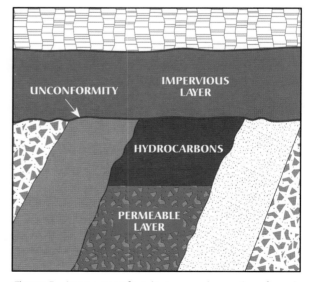

Figure 5. In an unconformity trap, an impervious layer is deposited on top of a hydrocarbon-bearing layer.

A PRIMER OF OFFSHORE OPERATIONS

To understand unconformity traps, think of several layers of rock being deposited horizontally over millions of years, somewhat like a giant layer cake. Then, for some reason, deposition stops. Maybe the sea in which the deposits are occurring evaporates or retreats. As more time passes, earth movements tilt the layers. As still more time passes, geologic movement exposes these tilted layers to weathering. Wind, water, or ice erode, or weather, the tilted layers. During this period, no more depositing occurs. Finally, another change occurs in the environment so that deposition starts again. New beds form on top of the old, weathered ones. The result is a series of beds, all in contact with one another, but with a time gap in the sequence of layers. The gap resulted because no rocks were deposited during a period of time in the history of the structure. Geologists call such interruptions in the continuity of deposition "unconformities."

DOMES OR PLUGS

Other types of traps where geologists may find oil and gas are *domes* or *plugs* of material, usually salt, that pierce and deform overlying strata. As the molten salt shoves its way upward through the rock layers, it tilts those layers it pierces and folds those above. Eventually, the molten salt cools and solidifies, leaving deformed rock structures around it that can serve as traps. Hydrocarbons can migrate into any porous and permeable beds around the column of salt and be trapped there, since salt is impermeable. The folded, porous beds above the salt can also serve as hydrocarbon traps if overlain by an impervious layer (fig. 6).

PERMEABILITY CHANGES

Some layers of rock do not have the same permeability throughout their structure. During the early stages of formation, perhaps the rock-making forces on sediments were relatively weak, thus creating rock that was not too dense. Later, the forces became higher for some reason and compacted the upper sediments very, very tightly, forming extremely dense rock. As a result, the lower sediments had relatively high permeability, while the upper sediments had little if any permeability. When hydrocarbons migrated into the lower and permeable sections of the bed, the upper and relatively impermeable sections prevented them from further movement (fig. 7).

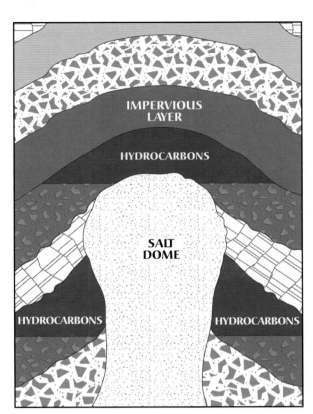

Figure 6.   Hydrocarbons can be trapped around and above a salt dome.

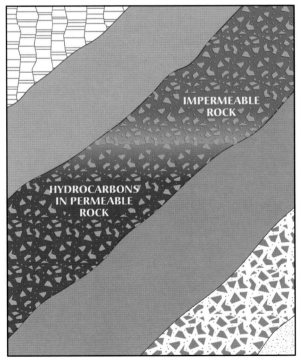

Figure 7.   A change in permeability within a single rock layer can trap hydrocarbons.

COMBINATION

A very common reservoir trap is one that a *combination* of folding, faulting, changes in permeability, and other factors create. For example, a faulted anticline associated with an unconformity can create a reservoir (fig. 8). Combination traps are the most common type of trap because the earth's great forces that deform rock layers usually do so in many different ways.

Knowing how oil and gas can be trapped in buried reservoirs is one thing, but finding these traps is quite another.

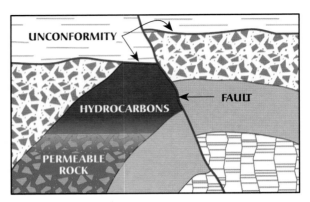

Figure 8.   A combination of folding, faulting, and an unconformity can trap hydrocarbons.

Summary

Oil and gas are hydrocarbons. Two elements make them up: hydrogen and carbon.

The simplest hydrocarbon is methane, CH_4; it is the main component of natural gas. Methane is an odorless, colorless gas that is lighter than air.

Crude oil is often called petroleum, which means "rock oil." Although crude oil varies widely in appearance, it is usually a liquid that weighs less (has a lower density) than water.

Geologists find oil and gas in the pores of buried layers of rock. If passages connect the pores of the rock, the rock is permeable. If a rock layer is permeable, any oil and gas can flow from the pores into a well. Oil and gas usually occur in sedimentary rocks. Sandstone and limestone are two common types of sedimentary rocks in which hydrocarbons occur.

The buried remains of microscopic plants and animals probably formed oil and gas in the distant past. Once formed, subsurface pressures forced the hydrocarbons from their place of origin and moved them upward. In most cases, underground formations trapped the hydrocarbons before they reached the surface.

Many different types of traps exist. In general, folding, faulting, unconformities, domes or plugs, changes in permeability, and combinations of these phenomena form traps.

2
EXPLORATION

In the early days of the industry, oil explorationists usually looked for oil springs and drilled their wells nearby. They soon learned to look for other surface features that might show the presence of a subsurface reservoir. For instance, they found that salt domes often created traps for oil and gas. So they looked for upward bulges in the surface that might indicate an underlying dome and drilled wells around it.

Today, however, the search is more difficult. The early oil operators quickly drilled all the relatively shallow traps that showed their presence on the surface. Deep reservoirs usually do not give any indication on the surface. Explorationists cannot therefore find them by direct observation. And offshore, where seas and oceans cover the seafloor, oil finders must depend entirely on indirect scientific methods—at least in the first stages of exploration.

Indirect methods of hunting for oil and gas depend on the fact that rocks have variable properties. For instance, different rocks have different magnetic properties, and some rocks are denser than others. Sensitive instruments can measure and record these properties and often locate subsurface formations that *may* contain hydrocarbons.

Emphasize the word "may," for even if an oil hunter finds a subsurface formation with the right shape or configuration, the formation may not have any hydrocarbons. The only way to find out for sure is to drill a well into it. In other words, indirect methods reveal only the possibility of a petroleum trap. To confirm that the trap holds hydrocarbons, oil companies use a direct method: they drill a well.

Before a company can begin drilling, however, the company's decisionmakers must determine where. They cannot afford to put a rig in an arbitrary spot; they have to have some indication that the spot holds promise. In the search for that spot, four steps may take place. First, the company may run a magnetic survey. Second, they may make a gravity survey. Third, they may carry out seismic surveys. Finally, if the survey results look promising, they may drill an exploratory well.

All four of these steps may occur in sequence, especially in prospecting for oil and gas on land. Offshore, however—where companies carry out most exploratory work from a boat carrying a large amount of equipment—they may run magnetic, gravity, and seismic surveys simultaneously.

Magnetic Surveys

To run a *magnetic survey*, the boat tows a magnetometer in the water over the area of investigation. The *magnetometer* is a very sensitive instrument. It measures and records the magnetic forces of the rocks in the earth's crust. If the rocks are homogeneous—all pretty much the same kind of rock—the magnetometer record shows a fairly uniform magnetic field. But the magnetometer may record a distortion in the magnetic field. This distortion, an *anomaly*, tells surveyors that they have found rocks containing minerals that attract a magnet. Finding such rocks may indicate an area of promise.

Basement rock, an igneous rock that in many places lies under sedimentary layers, often contains minerals that are magnetic. Basement rock seldom contains hydrocarbons. Sometimes, however, it pushes its way upward, or intrudes, into overlying sedimentary rock. An intrusion of basement rock may create arches and folds, or anticlines, which geologists also call *highs*, in the sedimentary rock. These highs may serve as hydrocarbon traps.

Magnetic surveying is not foolproof. Sometimes it cannot detect known traps. For example, in the Gulf of Mexico, salt domes intruded into the sedimentary rocks lying below the water and formed numerous traps. Explorationists discovered them without using magnetic surveys. Interestingly, a company later ran several magnetic surveys after it had drilled many successful wells in the area. The surveys failed to reveal the salt domes. Apparently, a magnetometer cannot detect the very small magnetic anomalies associated with salt domes. On the other hand, magnetic surveys work well where the intruded rock is granite or other igneous rock with strong magnetic properties.

As long as oil finders keep the limitations of magnetic surveys in mind, they can use them successfully to narrow the offshore search for oil and gas.

Gravity Surveys

In addition to or in place of magnetic surveys, the offshore oil prospector may use gravity surveys. A *gravity survey* involves the use of a very sensitive instrument—a *gravimeter*, or gravity meter—that survey crew members mount on a boat. Since the gravimeter is so sensitive, wave motion can make its readings unreliable. Therefore, crew members must mount it in such a way that it remains stable in spite of motion. Inside the gravimeter is a small metal object that a rock's density will affect. As the boat moves over the water, the gravimeter weighs the object. If the meter and object pass over dense rocks, the weight of the object increases. Conversely, if the meter and object pass over light rocks, the weight of the object decreases. The density of rocks in the subsurface alters the earth's gravitational pull. The altered gravitational pull affects the weight of the object in the meter. By measuring and recording the weight of the object in the meter, explorationists can get an idea of the density of the underlying rocks.

The density of rock formations indicates the possibility of traps. For example, intruded basement rock is usually denser than overlying sedimentary rock. Denser rock shows up as a *positive gravity anomaly* on the record made by a gravity survey. A positive anomaly may indicate an igneous uplift (fig. 9). Hydrocarbons sometimes accumulate in overlying sedimentary rocks that igneous rocks uplifted into arches or folds.

Some intrusions, like salt domes, are less dense than the overlying sedimentary rocks. In such cases, a gravity survey may indicate the less dense salt as a *negative gravity anomaly* (fig. 10).

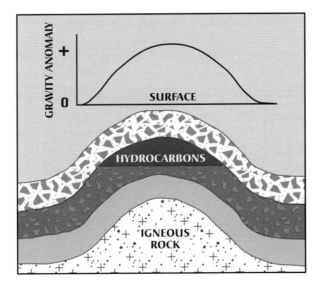

Figure 9. A positive gravity anomaly over an igneous uplift can indicate the presence of a hydrocarbon accumulation.

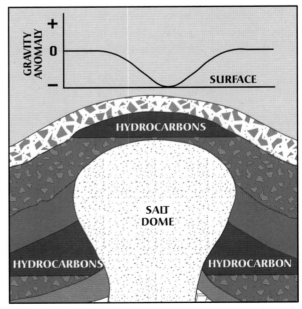

Figure 10. A negative gravity anomaly can indicate the presence of a salt dome with associated hydrocarbon accumulations.

Whether positive or negative, such gravity anomalies may indicate hydrocarbon traps.

Gravimeter surveys have been especially useful in pinpointing salt domes. However, like most exploration surveys, the results aren't infallible. For instance, salt isn't the only substance that shows up as a negative anomaly. Thick beds of low-density shale also give negative gravity readings. Unfortunately, lightweight rocks aren't usually associated with traps as often as salt domes are. Yet the gravity surveyor has no sure way of knowing whether the meter is indicating the presence of salt or a low-gravity rock. Nevertheless, in the hands of a skilled surveyor, gravity surveys help narrow the search for offshore petroleum reserves.

Seismic Surveys

Seismic surveying is a widely used indirect method of exploration, both on land and offshore. It provides a lot of detailed information about subsurface formations. As a result, seismic surveying very often supplements gravity and magnetic surveying. Gravity and magnetic surveys provide preliminary information; seismic surveys provide more precise detail. A seismic survey is usually the last exploration step before a company actually drills a prospect.

SEISMIC SURVEYING PRINCIPLES

Seismic surveying depends on the fact that rock layers reflect sound waves. That is, sound bounces off the layers like an echo. Rock layers also refract sound waves. That is, the rock layers deflect, or bend, the sound waves as they travel through the rock. Oil seekers don't, however, use sound refraction as much as they use sound reflection in modern seismic work.

To understand modern seismic surveying, remember that many kinds of rock are present in the subsurface. In general, the rocks occur in layers, one on top of the other, much like the layers of a cake. Offshore, a boat trails a special noisemaker that generates a loud, low-frequency sound. The loud, low-frequency sound waves go down through the water and into the rock layers. They then continue traveling downward through the many layers for thousands of feet (metres).

Where one kind of layer ends and another begins, the boundary reflects some of the sound waves. The reflected sound waves travel back up through the layers and into the water. In the water, sensitive detectors called *hydrophones*, which the boat also tows, pick up the sound waves. The reflections from shallower layers arrive at the hydrophones sooner than the reflections from deeper layers (fig. 11).

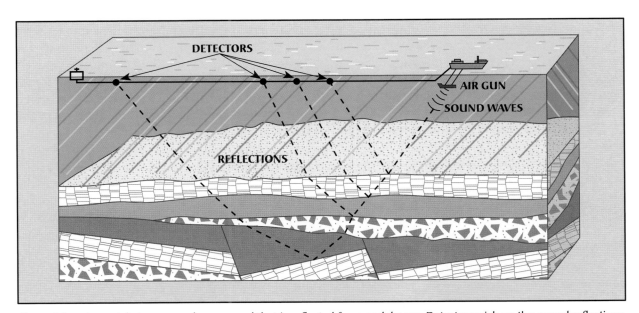

Figure 11.   A special air gun produces sound that is reflected from rock layers. Detectors pick up the sound reflections.

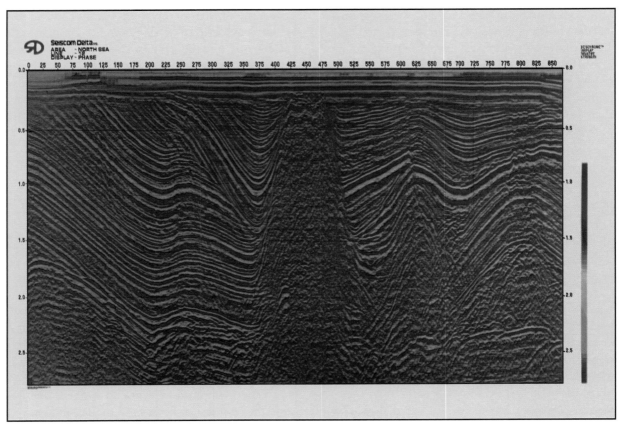

Figure 12. A seismic section, color-enhanced by a computer in this example, gives a view of the subsurface and thus the possibility of traps.

The arrival times of each reflection indicate the depth of the layers, giving explorationists a cross-sectional view of the subsurface. Geologists call this cross section a *seismic section*, or a *record section* (fig. 12).

Expert interpreters using powerful computers can look at a seismic section and tell if a geological structure might contain hydrocarbons. Seismic data do not directly indicate hydrocarbons—with one possible exception.

Sometimes natural gas shows up on a seismic section as a *bright spot* (fig. 13). Bright spots are areas of intense sound reflection. Unfortunately, other subsurface phenomena can also show up as bright spots. So it is still true that the only sure way to verify that hydrocarbons exist is to drill a well.

SEISMIC TECHNIQUES

In offshore seismic work, an underwater *sound generator* usually makes the sound waves. The boat trails the sound generator behind it as it moves over the area. The boat also trails several hydrophones.

At specific times, an automatic device sets off the sound generator. The generator suddenly releases compressed air, which makes a loud noise. (Explorationists used to set off dynamite to make the noise. Now they use compressed air because it does not harm marine life.) The sound travels downward through the water and enters the subsurface rock formations. The many rock layers then reflect the sound back into the water, where the hydrophones pick it up (see fig. 11).

The hydrophones convert the sound energy to electrical pulses. A cable sends the pulses back to the boat. There, a special recorder records the pulses from the hydrophones. After the geologists collect the seismic data, they send the data to a laboratory. In the lab, powerful computers process and enhance the data. The end result is a seismic section that experts can read. If the section shows that there is a good chance of a hydrocarbon-bearing trap, then an oil company may drill an exploratory well.

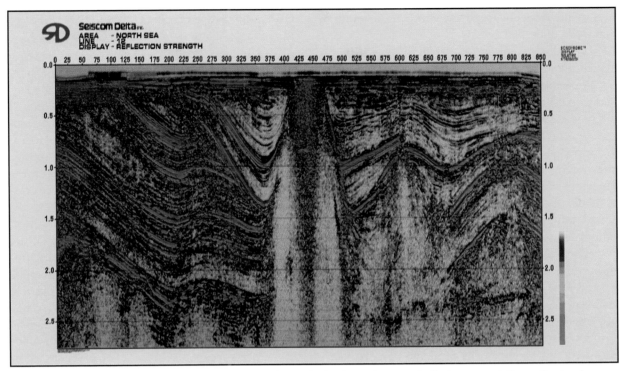

Figure 13. Bright spots, shown in red on this color-enhanced seismic section, may indicate accumulations of natural gas.

Survey Locations

Besides collecting the seismic data, the surveyors also record the exact location on the water's surface where they gathered it. If the data look promising, an oil company needs to know precisely where the seismic crew obtained them. They can then place a drilling unit in the most favorable location over the subsurface rock structure.

In some cases, *radio triangulation* finds and records the position. In this method, two shore-based radio stations communicate with a portable radio station on the boat. Both the shore-based stations and the portable station transmit and receive signals. The time it takes for signals to travel between the boat and each of the land stations indicates the precise distance between the boat and the two stations. With the two distances, a person on the boat can very accurately determine the boat's location.

In most cases, companies use a *geopositioning satellite (GPS) system*. In this technique, a sensitive transceiver on the survey boat and transceivers on orbiting space satellites precisely locate the position of the survey. The satellites receive and transmit signals to and from the GPS transceiver. The GPS then calculates and displays the boat's position by latitude and longitude to within a foot (a half metre) or so.

Obtaining Drilling Rights

An oil company that wishes to drill for and possibly produce hydrocarbons must obtain the rights to do so. In most cases, the company must obtain these rights from the country in whose waters the oil and gas activity will take place.

OBTAINING RIGHTS IN THE U.S.

In the United States, most offshore exploratory drilling occurs in an area known as the *Outer Continental Shelf*, or OCS. The Outer Continental Shelf starts at the point offshore where state ownership of the water and of any minerals (such as hydrocarbons) under the water ends.

This distance varies from state to state. For instance, the state of Texas owns the mineral rights from shore to a distance of 9 miles (14.4 kilometres). Louisiana, on the other hand, owns the mineral rights from shore to a distance of 3 miles (4.8 kilometres).

Beyond the area of state ownership in the Outer Continental Shelf, the federal government controls the waters and minerals. The Department of the Interior, Minerals Management Service (MMS), controls the OCS. The OCS ends where international waters begin or where treaties with other countries establish the end of U.S. jurisdiction.

Thus, to explore in OCS waters of the United States, the oil company or companies (sometimes several work together) must obtain the right to do so from the federal government. Companies usually obtain this right by bidding on offshore blocks, or tracts, offered for sale by the government at various times. The company or companies that bid the most money are the ones most likely to win the right to drill for and produce any hydrocarbons discovered on the block.

OBTAINING RIGHTS IN OTHER COUNTRIES

In countries other than the United States, the system is similar, except that most other nations own all the mineral rights from the shoreline out to international waters or the waters of bordering countries. For example, in the North Sea, which is bordered by the United Kingdom, Norway, Denmark, Germany, and the Netherlands, the area is divided into sectors. Each country controls its own sector. In this area, the country in whose sector the drilling site lies grants the oil companies concessions to drill.

As you might imagine, obtaining the right to drill for oil and gas and the right to sell any oil and gas discovered can be a complicated business. Drilling for hydrocarbons costs not only large amounts of money but also many hours of time and research. From the time explorationists make the first surveys to the time a company produces the first oil and gas from an offshore site—if they ever produce any—fifteen years may pass. Therefore, any companies that become involved in offshore ventures must be very careful in their survey work to ensure the best chance of success.

Summary

Oil companies use various kinds of surveys to find a site on which to drill an exploratory well. Magnetic surveys involve the use of a magnetometer, which measures the magnetic forces in the rocks lying in the earth's crust. By measuring and recording magnetic anomalies, explorationists can find arches and folds that might serve as hydrocarbon traps.

Gravity surveys measure and record the density of underlying rocks. Positive and negative gravity anomalies may indicate the presence of folds and domes that could serve as hydrocarbon traps.

Seismic surveys measure and record the time it takes for loud, low-frequency sound to bounce off subsurface rock layers and reach the surface. A seismic section is a cross-sectional view of the rock layers under investigation. Experts can find geological structures favorable to accumulations of hydrocarbons from seismic sections.

Even though surveys help pinpoint the existence of possible hydrocarbon traps, the only sure way to confirm their presence is to drill a well. However, before the company can drill a well, they must obtain the rights to do so from the country in whose waters the proposed drill site lies.

3

DRILLING RIGS

Once a company has obtained the right to drill a *wildcat*, or *exploratory*, *well* to see if hydrocarbons exist, they must then select some type of drilling *rig*. More often than not, they will use a *mobile offshore drilling unit* (*MODU*; pronounced "moe-doo") (fig. 14).

Rig owners can move mobile offshore drilling units from one drill site on the water to another. A rig has to be mobile because, after it finishes drilling one exploratory well, a crew has to move it to another site—perhaps nearby, perhaps far away—to drill another.

Oil operators use two basic types of MODUs to drill most offshore wildcat wells: bottom-supported units and floating units. Bottom-supported units include submersibles and jackups. Floating units include drill ships and semisubmersibles (fig. 15). Of the many types of MODUs, operators and contractors use jackups, semisubmersibles, and drill ships the most. Jackups are the most common.

Figure 14.   A mobile offshore drilling unit (MODU) *(Courtesy Sedco Forex Schlumberger)*

Figure 15.   Four types of mobile offshore drilling units: (A) jackup; (B) drill ship; (C) submersible; (D) semisubmersible

Bottom-Supported Units

Two types of bottom-supported mobile offshore drilling rigs are the submersible and the jackup. Submersible rigs include posted barge submersibles, bottle-type submersibles, and arctic submersibles. Jackups differ mainly in the design of their legs and the way in which the manufacturer mounts the rig on the barge hull of the unit.

Currently, operators and contractors use jackups to drill most offshore wells. Semisubmersibles run a distant second, while drill ships and other floating units come in third. Submersibles run far behind.

When a bottom-supported unit is on site and drilling a well, a part of its structure is in contact with the seafloor. Special legs or columns support the remainder of the rig above the water (fig. 16). A crew can, however, move the rig, because it can float (fig. 17).

POSTED BARGE SUBMERSIBLES

The first mobile offshore drilling unit built was a submersible. Constructed in 1948, it drilled its first well in the Gulf of Mexico in 1949. It was near the mouth of the Mississippi River. It sat in 18 feet (5.5 metres) of water.

Figure 17.   Four boats tow a jackup as it floats on the water's surface.

The rig was a rectangular steel box—a flat-bottomed and flat-sided barge hull. Numerous steel posts rose from the top of the barge hull. The fabricators built a deck on top of the posts. They then placed the drilling equipment on the deck. Two pontoons, hinged and attached with cable, rode on both

Figure 16.   This jackup working in Indonesian waters is one type of bottom-supported unit. Each of its legs is in contact with the seafloor. *(Courtesy Marathon Oil Co.)*

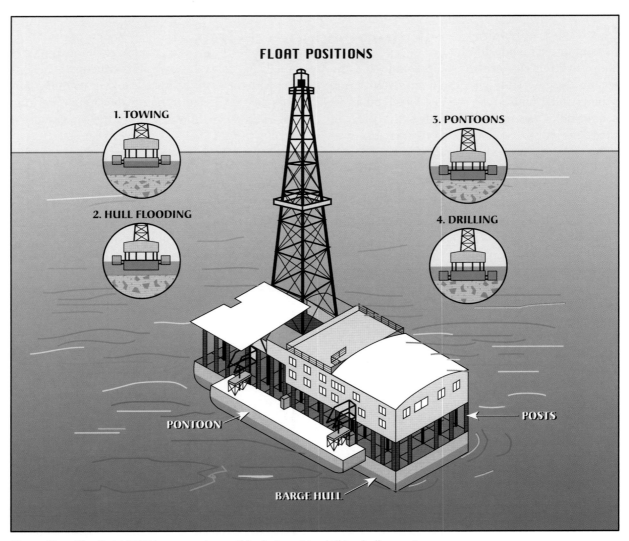

FLOAT POSITIONS

1. TOWING

2. HULL FLOODING

3. PONTOONS

4. DRILLING

PONTOON

POSTS

BARGE HULL

Figure 18.   The first MODU was a submersible designed to drill in shallow waters.

sides of the barge hull. These pontoons stabilized the unit (fig. 18).

When a towboat moved the unit, a crew emptied the water from the barge and the pontoons, which allowed the unit to float (fig. 18, position 1). A boat towed it to the drilling site and crew members flooded the barge hull; that is, they allowed water to enter the hull at a controlled rate (position 2). As water entered, the hull slowly settled to bottom, eventually coming to rest on the seafloor. Once the main hull rested on bottom, crew members flooded the pontoons (position 3), and they too settled to bottom (position 4).

The posts, since they extended some distance above the barge hull, allowed the drilling deck to stay above the waterline. Surface wave motion did not have much effect on the unit because the posts

were relatively thin and as such were fairly unaffected by wave action. The design worked well, and the entire structure provided a very stable platform from which the rig could drill.

These submersibles, often called *posted barge rigs*, worked so well in shallow water that manufacturers built quite a few of them. However, later versions did not use pontoons. In general, companies used them to drill wells in waters no deeper than about 30–35 feet (9–11 metres).

BOTTLE-TYPE SUBMERSIBLES

As drilling activity increased in the Gulf of Mexico, exploration surveys indicated several potential hydrocarbon reservoirs in the area. Since they lay offshore in waters too deep for posted barge submersibles to handle, in about 1954, rig designers came up with a new submersible, a *bottle-type submersible*, that could work in deeper water. Instead of a barge hull and pontoons, rig builders constructed the new design from several steel cylinders. They laid some of the cylinders horizontally; others they laid vertically, and then welded them together to form a sort of open cube (fig. 19).

Figure 19.   When flooded, the bottles cause a bottle-type submersible to submerge to the seafloor.

The builders then placed a cylinder much larger in diameter than the rest at each corner of the cube. They called these four large-diameter cylinders bottles. When flooded, the bottles caused the rig to submerge. When crew members removed the water, the rig floated.

The earliest submersibles of this type could drill in waters up to 70 feet (21 metres) deep. Later designs drilled in water depths up to 100 feet (about 30 metres). One bottle-type submersible built in 1962 (a triangular shape) could work in water as deep as 175 feet (53 metres).

ARCTIC SUBMERSIBLES

The latest submersible rigs are *arctic submersibles*. In the arctic, open, ice-free water exists only for short periods in the summer. So, the rig owner has to wait for the thaw, then quickly move the rig into position and submerge it before the sea freezes up. While drilling, arctic submersibles can withstand the tremendous forces of moving pack ice that surrounds them most of the year.

One arctic design is a *conical drilling unit*, or *CDU*. Another is a *mobile arctic caisson*, or *MAC*. Still another is a *concrete island drilling system*, or *CIDS* (fig. 20). Heavy steel or concrete walls—a *caisson*—surround the equipment below the waterline. The caisson protects the equipment from damage by moving ice. The drilling deck, or platform, has no square corners so that moving ice can better flow around it.

Figure 20.   A concrete island drilling system (CIDS) features a reinforced concrete caisson with a steel base.

JACKUPS

The industry built the first *jackup*, or *self-elevating*, *rig* in 1954. It rapidly became the most popular design in mobile offshore drilling units. Jackups are popular because they provide a very stable drilling platform, since part of their structure is in firm contact with the bottom of the ocean. They can also drill in relatively deep water (the biggest can drill in waters about 350 feet or 107 metres deep). What is more, towboats can easily move a jackup from one location to another.

An offshore drilling contractor can choose from two basic types of jackups. One has open-truss legs. In this design, manufacturers construct the legs from tubular steel members. They then crisscross the members to make very strong, but relatively light-weight, legs. Open-truss legs look somewhat like the towers that carry high-voltage electric lines across the countryside (fig. 21A).

The other jackup has columnar legs. Columnar legs are big steel tubes (fig. 21B). Columnar legs are less expensive to build than open-truss legs. They cannot, however, withstand bending stresses as well as open-truss legs. As a result, even the largest columnar-leg jackups cannot drill in waters much over 250 feet (75 metres) deep.

Whether columnar or truss type, the legs of a jackup pass through openings in a barge hull. Three or four legs are common, but some designs call for more or fewer. The deck of the barge serves as the platform for drilling equipment and other machinery.

Moving a Jackup

When a crew moves a jackup, it raises, or jacks, the legs up out of the water so that the barge floats. With the legs completely out of the water, the rig movers can transport the entire rig to the drilling location (see fig. 17).

Usually rig movers tow the rig, but at least one jackup is self-propelled. It has two engine-driven

Figure 21.   Jackups may have either open-truss legs (A) or columnar legs (B). *(Courtesy Sundowner Offshore Services, a Nabors Industries Company)*

A PRIMER OF OFFSHORE OPERATIONS

Figure 22.   A special ship carries a semisubmersible rig. *(Courtesy Sedco Forex Schlumberger)*

propellers, or screws, mounted on the unit that can move it.

Sometimes, large ships with flat decks move jackups. Crew members submerge the ship so that its deck is below the water. They then maneuver the rig onto the submerged ship's deck. With the rig in place, crew members pump out the water from the ship, which allows it to float. Especially for long-distance moves, ships can carry the jackup at speeds faster than boats can tow it. Ships can also carry semisubmersibles (fig. 22). As you can imagine, a ship used for this job is very large.

Jacking Devices
Once the jackup rig is on location, crew members jack the legs down until they come into firm contact with the seafloor. Then they jack up the barge hull onto the legs until it is clear of the water's surface and well above high waves (fig. 23).

One of two systems raise or lower the legs and barge hull. One uses hydraulic cylinders fitted with moving and stationary pins. The cylinders extend and retract to climb up or down the legs, in much

Figure 23.   A mobile offshore drilling unit (MODU) *(Courtesy Sedco Forex Schlumberger)*

the same fashion as you would climb a rope. The second uses a rack-and-pinion system. In this system, a series of round gears—pinions—engage a flat, long gear—the rack. The rack normally forms

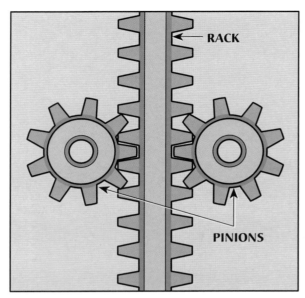

Figure 24.   A rack and pinion consists of a rack and two pinions that mesh with the rack.

the corners of each leg (fig. 24). Large electric or hydraulic motors turn the pinion gears. Manufacturers mount the pinions on the deck of the barge hull. When the pinion gears turn, the racks, or legs, move up and down (fig. 25).

Figure 25.   A rack on each leg of a jackup allows the pinions to raise or lower the leg.

Bottom Support for Legs

Large mats support some jackups while spud cans support others. On one type of *mat-supported jackup*, the rig builders attach the bottom of each leg to a large, generally **A**-shaped steel frame (fig. 26). In general, companies use mat-supported units where the sea bottom is soft and muddy. The mat distributes the weight of the rig evenly over the bottom

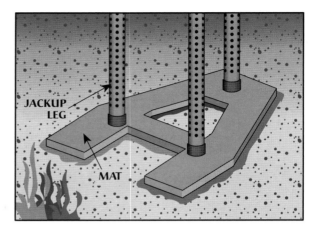

Figure 26.   The mat on a mat-supported jackup rests on the seafloor and prevents the legs from penetrating the ooze.

and tends to keep it from sinking too far into the ooze. The mat works much like a stepping stone in a muddy patch of ground; when you step on the stone, its broad support keeps you from sinking ankle-deep in mud.

Manufacturers put spud cans on independent-leg jackups. (The legs are independent because a mat or other device does not join them.) *Spud cans* are steel cylinders, often with pointed ends, that the builder attaches to the bottom of each leg (fig. 27). Where the seafloor is not too soft and muddy, the spud can on each leg penetrates into the bottom a short distance and stabilizes the rig. A spud can works like the sharpened end of a pencil jammed

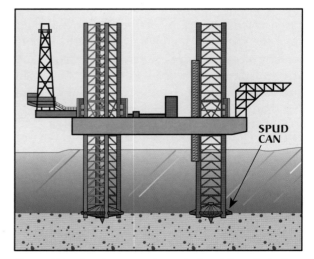

Figure 27.   Spud cans on the end of each leg of a jackup penetrate a short distance into the seabed and stabilize the rig.

into the ground. If you jam it down hard enough, the ground firmly supports the pencil.

When drilling with independent-leg jackups in areas where the operator and contractor are not familiar with the seafloor's composition, engineers carefully test its strength. It has to be strong enough to support the great weight of the jackup. Otherwise, one or more of the legs could punch through the seabed and the rig could tilt or capsize.

Derrick Package Mounting

Rig builders can mount the derrick package in different ways on jackups. In one instance, they mount the derrick on two very heavy steel arms that protrude outward from the platform deck. This is a *cantilevered jackup* (fig. 28).

In another instance, the builder mounts the derrick over a slot—a *drilling slot*, or a *keyway*—that is in the edge of the platform deck (fig. 29). Because the cantilevered design is more versatile, cantilevered jackups have replaced almost all those with keyways.

Figure 28. The derrick package on this cantilevered jackup is supported by large steel beams that extend from the platform deck.

Figure 29. The derrick on this jackup rests over a slot (keyway) in the platform deck.

Floating Units

Floating rigs include drill ships and semisubmersibles. Generally, rig owners and operators most often use drill ships to drill wells in the deep, remote waters of the world. Semisubmersibles can drill in seas too rough for drill ships.

Drill ships have a ship-shaped hull that allows the drilling contractor to store a lot of supplies and equipment on board the vessel. Thus, a drill ship can stock up in port, sail to the drilling location, and drill for long periods without being restocked.

Semisubmersibles have large pontoon hulls and big columns, which the rig operator submerges to a depth below the water's surface. Submergence makes semisubmersibles the most stable of floating drilling units. Even though semisubmersibles cannot carry as much drilling supplies as a drill ship, contractors often have to use them to drill in rough seas.

DRILL SHIPS

Even though drill ships strongly resemble passenger or cargo ships, the differences between them are very easy to spot. First, a drill ship has a drilling derrick, a tall, towerlike structure necessary for any drilling operation. Rig builders usually locate the derrick *amidships*, a nautical term that refers to the middle of the ship (fig. 30).

Figure 30.   A drill ship *(Courtesy Sedco Forex Schlumberger)*

In addition, most drill ships have a *moon pool*, a walled opening below the derrick, open to the water's surface and through which various drilling tools can pass down to the seafloor.

And finally, most drill ships have a helipad, which the rig builder usually locates at one end of the vessel. Helicopters land on the pad, usually when letting off or picking up rig crew members or other personnel.

Early Drill Ships

Marine architects developed drill ships about the same time as posted barge submersibles, in the late 1940s. Posted barges worked well in the shallow waters of the Gulf of Mexico. Oil explorers and producers working off the Pacific coast of California were, however, faced with a problem that submersibles couldn't handle: the water was too deep.

The continental shelf off the California coast is an underwater area that extends from the shoreline to a water depth of about 600 feet (about 180 metres). At this depth, the continental slope begins. (This continental shelf is not the legal Outer Continental Shelf, or OCS, of the United States. All continents have a continental shelf. However, congress designated the Outer Continental Shelf as an area of government jurisdiction in the United States.)

The geographic shelf off California was a problem for oil explorers and producers, since in many places it is only a few miles (kilometres) wide. Outward from the shelf, where exploration surveys indicated the possibility of hydrocarbon deposits, the water was too deep for posted barge submersibles. So designers had to develop some sort of floating drilling platform. They turned to ships for a solution.

The first vessel they used was a surplus U.S. Navy patrol craft. The builders erected a small derrick on a cantilevered platform. They suspended the platform off the port (left) side of the ship. They successfully drilled enough shallow holes with the ship to encourage the companies involved in the project to go a step farther.

The companies obtained a surplus ship, cut a walled round hole, or moon pool, in the center of it, erected a derrick over the moon pool, and began drilling a series of wells in deep water. So successful was this ship that the companies soon ordered the construction of the first ship built exclusively for offshore drilling.

Utilization of Drill Ships

Drill ships are the most mobile of all the mobile drilling units and are, therefore, often used in remote waters. Drill ships can drill wells in very deep water where bottom-supported units are not practical. However, they have one major drawback: because they float, they are very susceptible to wave motion and are not suitable for use in heavy seas.

Fortunately, since many waters of the world enjoy relatively pleasant seas most of the year, you can find drill ships working in many areas. One possible exception is in the North Sea, where operators consider seas running 20 feet (6 metres) high to be relatively calm.

Dynamic Positioning and Spread Mooring

Once a drill ship is on the drill site, the rig operator can keep it there in two ways. An anchor crew can deploy several anchors from the vessel to the seafloor, much as any large ship anchors offshore.

Or a rig builder can install several *thrusters* (powered propellers) fore and aft (front and back) and on both sides of the vessel. A computerized system automatically actuates the thrusters to maintain the vessel precisely on station. Special sensors on the well and on the drill ship constantly read the ship's position in relation to the wellhead. The sensors send signals to a computer on board the drill ship. The computer monitors the rig's position. It controls (starts and stops) the thrusters' action to offset wind, waves, and currents, all of which try to move the rig out of position. Such a mooring system is called a *dynamic positioning system* (fig. 31). In general, rig owners use dynamic positioning where the water is so deep that conventional anchors cannot do the job.

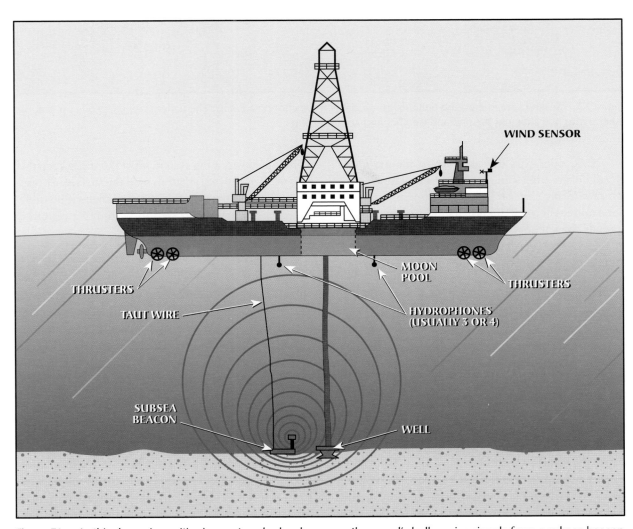

Figure 31. In this dynamic positioning system, hydrophones on the vessel's hull receive signals from a subsea beacon. The signals and information from a taut wire and a wind sensor are transmitted to onboard computers. The computers process all data and activate the thrusters to maintain the floating unit on station.

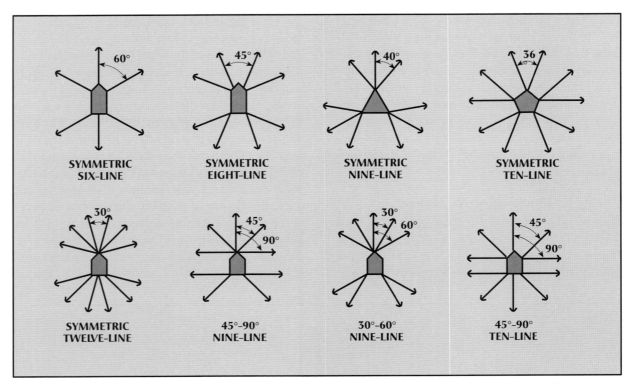

Figure 32.   Several spread mooring patterns are available for anchoring a floater. The pattern used depends on, among other things, the shape of the unit and weather and sea conditions.

If rig operators use anchors to maintain the vessel on site, they can employ one of a number of *spread mooring patterns* (fig. 32). Mooring patterns determine where an anchor handling crew sets or spreads the anchors and their lines on the seafloor in relation to the drilling vessel. Typically, crew members set eight or ten anchor lines at various points around the ship to keep it on station.

Whether rig operators use dynamic positioning, anchors, or both to maintain the vessel on site, they must be precise. The actual well is a relatively small-diameter hole the rig is drilling in the seafloor to some predetermined depth, usually thousands of feet (metres) deep. Further, the vessel itself may be riding the waves hundreds—perhaps even thousands—of feet (metres) above the seafloor. Stretched between the small well opening on the ocean floor and the vessel on the surface is tubular equipment that attaches the vessel to the well.

The equipment is flexible to allow the drilling vessel to move up and down with the waves, and sideways, frontwards or backwards with wind, waves, and currents. However, the movements cannot flex the tubular equipment beyond its design limits; otherwise the equipment breaks. The anchors or dynamic positioning system must therefore keep the vessel within a very small area on the surface. Spread mooring and dynamic positioning make it possible to keep the vessel on station within the required limits of movement.

Motion Factors

Drill ships are floaters, so rig operators always have to contend with the motion of the sea, whether small or large. Motion is especially troublesome when the vessel is on location and drilling. Over the years, engineers have developed special equipment to offset motion effects on the drilling operation.

One problem has been how to connect the top of the stationary well with the equipment on the moving drilling deck of the vessel. The fact that the well may be on the seafloor in thousands of feet (metres) of water complicates the problem. The solution lies in using *riser pipe*—special pipe that extends the top of the well from the bottom up to a position just below the drill floor of the vessel (fig. 33) and in using special devices that compensate for rig motion.

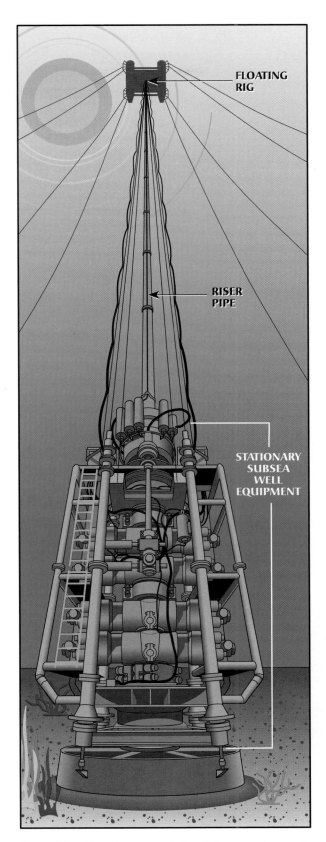

Figure 33. Riser pipe connects the stationary subsea well equipment to the surface equipment on a floating rig.

SEMISUBMERSIBLES

Ship-shaped floating units are very susceptible to wave motion. Operators can, however, find oil and gas in formations that lie beneath waters that are rough and stay rough most of the time. Therefore, naval architects had to come up with a floating rig that could withstand rough seas. Two designs evolved: bottle-type and column-stabilized semisubmersibles.

Bottle-Type Semisubmersibles

In their search for a rig design that could withstand rough seas, naval architects first turned their attention to bottle-type submersible rigs. The architects speculated that these submersibles had potential.

With this type of submersible, crew members flood the bottles so that the rig submerges and comes to rest on the seafloor. The naval architects found that if they did not completely flood the bottles—that is, if they left some buoyancy—the rig would settle below the water's surface but still remain well above the seafloor. Operators could then moor the unit on station with anchors, which provided the only contact the rig had with the seafloor (fig. 34).

Figure 34. On a bottle-type semisubmersible, partially flooded bottles submerge the rig below the water's surface but above the seafloor.

In this semisubmerged state, the rig was not as susceptible as a surface unit to wave motion, particularly rolling and pitching. *Rolling* is the tendency of a floating object to roll about a horizontal line drawn from its bow to its stern. When the starboard (right) side of the vessel goes down, the port (left) side comes up, and vice versa. *Pitching* is the tendency of a floating object to move up and down from bow to stern. When the bow goes down, the stern comes up, and vice versa.

A semisubmerged rig is very stable, which is just what operators need for drilling in rough waters. So successful were semisubmerged bottle-types that drilling contractors still use them as either submersibles or semisubmersibles, depending on the environmental requirements. Once partial submersion proved to work, however, naval architects soon designed units that were exclusively semisubmersible (fig. 35).

Figure 35.   A semisubmersible at work *(Courtesy Sedco Forex Schlumberger)*

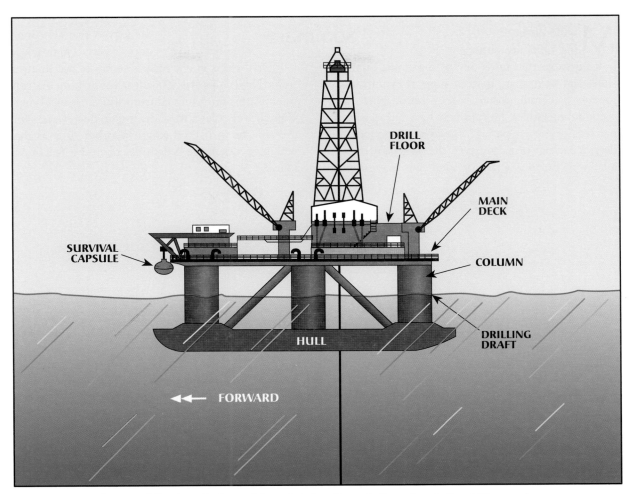

Figure 36. A column–stabilized semisubmersible rests below the waterline when in the drilling mode.

Column-Stabilized Semisubmersibles

One of the more popular semisubmersible designs is the *column-stabilized semisubmersible* (fig. 36). It consists of two longitudinal and streamlined lower hulls. The rig builders then mount several vertical cylinders or long rectangular cubes (columns) on the hulls. The columns, either round or square in cross section, extend upward from the hulls. The builders place the main deck on top of the columns.

Such semisubmersibles are often called column-stabilized units. Smaller diagonals crisscross between the upper deck and the large columns to give structural strength. At the drilling site, crew members flood the hulls to obtain the required drilling draft—the depth at which the rig rests below the water's surface. In most cases, anchors and chain moor the unit on station, and the rig owner uses spread patterns similar to those used for a drill ship (see fig. 32). Some semis use dynamic positioning

to keep them on station. (Offshore people like to shorten semisubmersible to *semi*.).

Since a lot of the mass of a semisubmersible submerges below the water's surface, the unit does not roll or pitch much. It does, however, tend to heave—move up and down. Naval architects can minimize heave tendencies without sacrificing too much stability using good design techniques. As on drill ships, the rig owner needs special equipment on a semi to compensate for heave.

When the contractor needs to move the rig, crew members pump water out of the hulls, and the entire unit floats on the surface. Once afloat, a rig moving company tows the rig to the next location. The ship-shaped lower hulls present a fairly streamlined surface to the water and thus make towing relatively easy. (Some semisubmersibles are self-propelled.) On long moves, the contractor may hire a special ship to carry the rig (see fig. 22).

Summary

Mobile offshore drilling units are either bottom-supported units or floating units. Submersibles and jackups are bottom-supported units. Submersibles include posted barges, bottle types, and arctic submersibles. Jackups may have either columnar legs or truss-type legs. Builders can mount the derrick package on a cantilever (the usual case) or over a keyway. Further, a mat or spud cans may support the legs. A mat almost always supports columnar legs.

Drill ships and semisubmersibles are floating units. Drill ships are best suited for drilling in deep, open waters far removed from shore. Semisubmersibles are also capable of drilling in deep waters but are able to withstand rough weather such as that which occurs in the North Sea or in the North Atlantic.

4
DRILLING A WELL

A writer once described a drilling rig as a portable hole factory. He pointed out that the sole purpose of a rig was to make holes in the ground and, since drilling contractors had to drill holes at different locations, the rig had to be movable. He wasn't too far off the mark. A rig, be it large or small, on land or offshore, has one main job: drilling wells.

Bits and Drilling Fluid

The job of a rig and the people who work on it is to put a drill bit in the earth and turn it. A *bit* is a hole-boring device (fig. 37). Pressed very hard against the ground and turned, or rotated, it makes a hole. The teeth on the bit grind and gouge the rock into small pieces. These pieces of rock, or cuttings, must be moved out of the way so the bit teeth can be constantly exposed to uncut rock.

As long as the cuttings are moved out of the way, the bit can drill ahead. To move cuttings away from the bit, the rig pumps a special liquid called *drilling fluid*, or *mud* (fig. 38). Most often, drilling

Figure 37.   A drill bit has teeth to penetrate a formation to make hole.

Figure 38.   Drilling mud is a special liquid essential to the drilling process.

mud is a mixture of water, special clays, and certain minerals and chemicals. Crew members very carefully mix drilling fluid to ensure that it can do its job well.

Special openings in the bit called *nozzles* eject drilling mud out of the bit with great speed and pressure. In fact, the mud actually jets out (fig. 39). These jets of mud lift the cuttings off the bottom of the hole and away from the bit. With the cuttings out of the way, they do not interfere with the bit teeth's ability to drill. Mud also carries the cuttings up the hole and to the surface for disposal.

Figure 39.   Drilling mud jets out of the bit nozzles to lift cuttings off the bottom of the hole.

Circulating System

The mud pump in the *circulating system* (fig. 40) (1) moves drilling mud down the hole through special pipe called *drill pipe* and *drill collars*, which are called the drill string; (2) mud then shoots out of the bit through jets and picks up the cuttings; and (3) carries them to the surface.

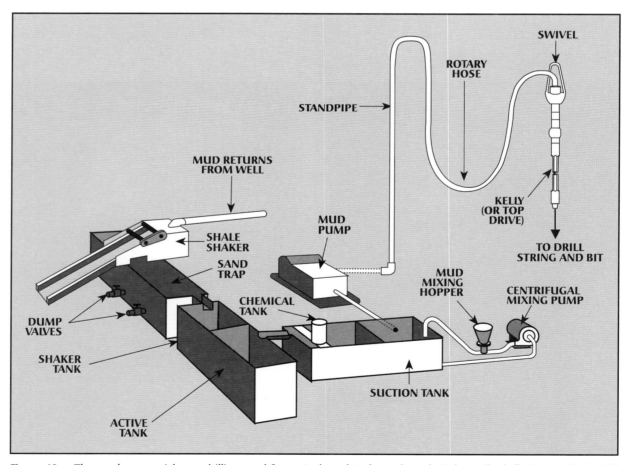

Figure 40.   The mud pump picks up drilling mud from steel mud tanks and sends it down the kelly (or topdrive), drill pipe, drill collars, and bit. Mud and cuttings return to the surface up the annulus.

MUD PUMPS

Large, heavy-duty pumps, the *mud pumps*, are the heart of the circulating system (fig. 41). (They pick up mud from rectangular steel tanks on the rig.) The pumps then force the mud into and down the drill pipe and drill collars and to the bit. At the bit, the mud jets out of the bit nozzles to move cuttings away from the bit. The mud then moves back up the hole to the surface.

MUD RETURN TO THE SURFACE

Since mud picks up cuttings made by the bit, it carries them as it returns to the surface. The mud and cuttings return to the surface in the space between the outside of the drill pipe and the inside of the hole. This space is the *annulus* (fig. 42).

At the surface, the mud and cuttings leave the well through a side outlet with a pipe on it. The pipe is the *mud return line*. At the end of the return

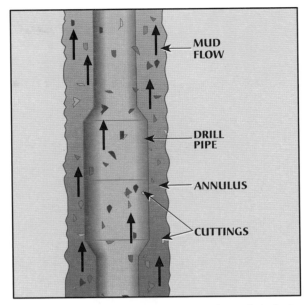

Figure 42.   Mud and cuttings return to the surface via the annulus, the space between the pipe and sides of the hole.

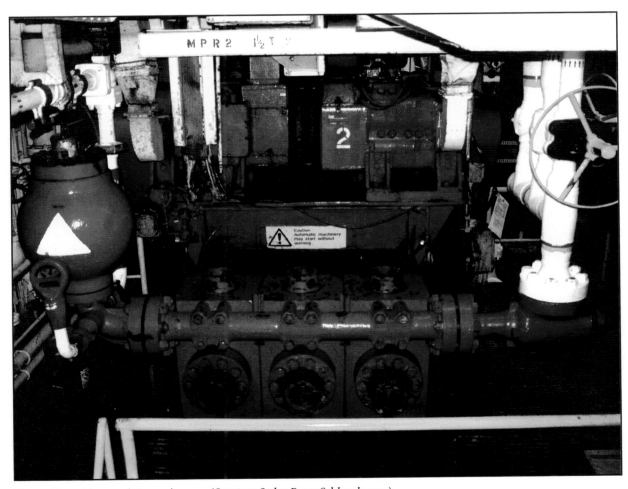

Figure 41.   A heavy-duty mud pump *(Courtesy Sedco Forex Schlumberger)*

Figure 43. A shale shaker screens drilled cuttings from the mud. *(Courtesy Sedco Forex Schlumberger)*

line, mud and cuttings fall onto a vibrating screen, or sieve, the *shale shaker* (fig. 43).

The shaker screens out cuttings but allows mud to pass through and fall into the mud tanks. Small-sized solids also pass into the mud tanks with the mud. Special equipment removes these solids. The equipment removes the solids to keep the mud at the proper weight (*density*). Mud that is too heavy or dense slows down the drilling process and holds down the cuttings on the bottom. The jets of mud cannot efficiently remove the cuttings, so the bit redrills old cuttings instead of making new ones from fresh, uncut formation.

Rigs use desanders, desilters, mud cleaners, and centrifuges to remove solids from mud. Desanders, desilters, and mud cleaners are *hydrocyclones* (fig. 44). Typically, hydrocyclones consist of several cones. The mud enters the cones and goes into a swirling motion, much like a tornado. This motion forces the solids to the side of the cone. The clean mud remains in the center and falls into the mud tanks.

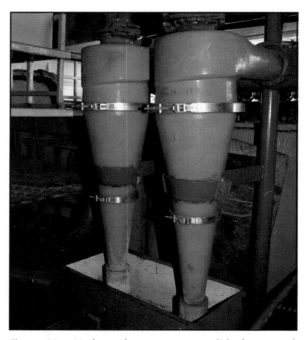

Figure 44. Hydrocyclones remove solids from mud. *(Courtesy Sundowner Offshore Services, a Nabors Industries Company)*

A PRIMER OF OFFSHORE OPERATIONS

Figure 45. A centrifuge spins mud at high speeds to remove solids. *(Courtesy Sedco Forex Schlumberger)*

Centrifuges spin the mud at a high rate of speed (fig. 45). Centrifugal force moves the solids to the outside of the centrifuge. Clean mud stays in the middle and moves into the tanks.

Finally, the pumps once again pick up clean mud and send it back down the hole. This circulating process goes on uninterrupted as long as the bit is on the bottom of the hole and drilling.

Rotating Systems

Part of keeping the bit drilling involves rotating it, so the rig needs a rotating system. Commonly, offshore rigs use a top drive (also called a power swivel) to rotate the bit. Another way rigs rotate the bit is with a rotary table and kelly system. And, in special situations, rigs can also use downhole motors to rotate the bit.

TOP DRIVE (POWER SWIVEL)

Most offshore rigs use a *top drive*, or power swivel, to rotate the drill string and bit (fig. 46). The rig's traveling block suspends the top drive. Top drives have a very powerful built-in electric or hydraulic motor (some have more than one motor). The motor or motors rotate a drive shaft. Crew members make up the top of the drill string to the drive shaft.

When the driller activates the top drive's motor and drive shaft, the drive shaft rotates the attached drill string and bit.

ROTARY TABLE AND KELLY SYSTEM

The *rotary table* and kelly rotating system is the traditional way rigs rotate the bit. Some rigs still use a rotary table and kelly system.

Kelly, Kelly Drive Bushing, and Swivel

A *kelly* is a special flat-sided length of pipe. It is usually six-sided, or hexagonal, in shape—that is, most kellys have a hexagonal cross section. (Some, however, are four-sided or square. Rig owners may use square kellys on smaller rigs, because a square kelly is not as strong as a hexagonal kelly.)

Figure 46.   A top drive (power swivel) is used to rotate the bit. *(Courtesy of National Oilwell)*

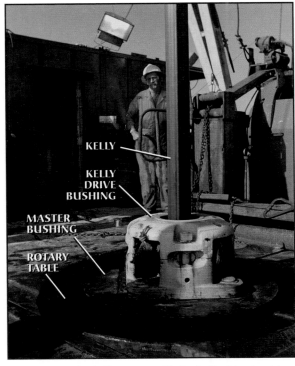

Figure 47.   The kelly mates with the kelly drive bushing, which fits into the master bushing of the rotary table. *(Courtesy Nabors Drilling, USA)*

Crew members make up the drill string to the bottom of the kelly.

The kelly fits into the *kelly drive bushing* (fig. 47). Flats inside the kelly drive bushing mate with the flat sides of the kelly.

Crew members make up the top of the kelly to a fitting on the *swivel* (fig. 48). The swivel supports the weight of the rotating drill string and allows drilling mud to pass into the drill string.

Master Bushing and Rotary Table
The kelly drive bushing fits into a master bushing. The *master bushing* fits inside the rotary table. When the driller activates the rotary table, it turns the master bushing. Since the kelly drive bushing fits into the master bushing, the master bushing rotates the kelly drive bushing and the kelly. Finally, since crew members make up the kelly on the drill pipe, the drill pipe, drill collars, and bit also turn. As the bit drills and deepens the hole, the kelly can move down against the flats in the kelly drive bushing.

To keep all this straight, think of it this way. The rotary table turns. The master bushing trans-

Figure 48.   The kelly is made up on the swivel. *(Courtesy Nabors Drilling, USA)*

fers this turning motion to the kelly drive bushing. And the kelly drive bushing transfers the turning motion to the kelly, drill pipe, drill collars, and bit. That's really all there is to it.

DOWNHOLE MUD MOTOR
Another way for the rig to rotate the bit involves drilling horizontal wells. Oil company engineers and technicians have learned that they can produce some reservoirs better if they drill a horizontal (instead of a vertical) well through the reservoir. In *horizontal drilling*, crew members begin by drilling a normal, vertical hole to a given depth. This depth is usually high above the reservoir's depth. They then deflect the hole from vertical. Over a distance of

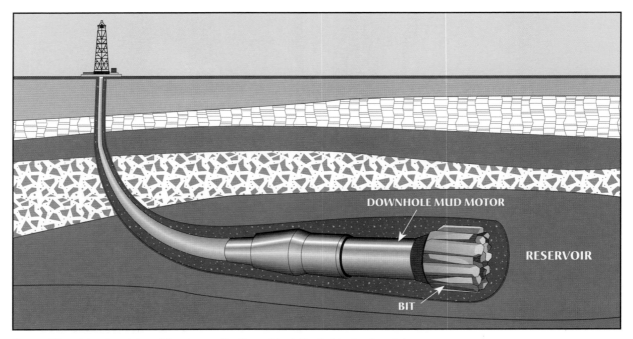

Figure 49. A horizontal well begins vertically and is deflected to horizontal.

perhaps hundreds of feet or metres, crew members curve the vertical hole until it becomes horizontal. The hole actually runs parallel to the surface rather than perpendicular to it (fig. 49).

When drilling a horizontal well, the rig crew cannot use a top drive or a conventional rotary table to rotate the drill string and bit. The reason has to do with the way in which they deflect the hole from vertical. Usually, crew members run a device—a *bent sub*—with a slight bend in it. This bend, usually from 1 to 3 degrees, points the drill string off to one side of the vertical hole (fig. 50). Crew members carefully point (orient) the bend in the string to make the bit go in the right direction.

Figure 50. A bent sub has a 1° to 3° bend in it to begin deflecting the hole from vertical.

While 1 to 3 degrees is not very much from vertical, the bend is enough to start the horizontal hole. Once started, crew members use special techniques to increase the curve more and more until the hole becomes horizontal.

The drill string now has a bend in it (drill pipe is flexible—it can bend a lot without breaking or permanently bending). The bend is also pointing the bit in the direction it needs to go. Crew members therefore cannot rotate the entire drill string. If they did, they would get a wobbling motion, rather than the straight rotating motion they get with a vertical string. Also, the bit would not drill in the right direction.

Since the rig cannot rotate the entire drill string when drilling a horizontal well, crew members install a downhole motor. A *downhole motor* is a strong length of pipe, inside of which is a special device that rotates the bit (fig. 51). Crew members attach the motor to the bottom of the drill string, and attach the bit to the bottom of the motor. With the bit and motor oriented in the right direction, the mud pump forces mud down the drill string and to the motor. In the motor, the mud strikes either a spiral-shaped shaft or several turbine blades. The great hydraulic force of the mud turns the shaft or blades. The shaft or blades (turbines) then turn the attached bit. Only the bit turns.

Figure 51. A downhole motor laid out on a land rig's pipe rack

Power System

Any drilling rig needs power—power to actuate the mud pumps, to hoist the drill string, and to run all the machinery on the rig. Several large diesel engines usually provide the power (fig. 52). A diesel engine is an *internal-combustion engine*, which means that it runs because a mixture of fuel and air burns inside the engine.

ENGINES

Some internal-combustion engines are spark-ignition engines, like the one you probably have in your car.

Figure 52. Several large diesel engines provide power for an offshore rig.

In these engines, spark plugs set off, or ignite, the fuel-air mixture. Other internal-combustion engines, however, are compression-ignition engines. In compression-ignition engines, like diesel engines, heat generated by compressing the fuel-air mixture also ignites the mixture. Anytime a gas like air is compressed, it heats up. When compressed enough, and a flammable substance like diesel fuel is injected, the air gets hot enough to ignite the fuel. Ignited fuel is the source of energy on which the engine runs.

Diesel engines power most offshore rigs. Usually, rigs require three or four engines. One by itself is not powerful enough—it doesn't have enough horsepower (kilowatts)—for all the power needs of the rig.

Rig engines, like almost everything else on a rig, are large and powerful. For the purposes of efficiency and cost effectiveness, engines are usually mounted on a base or skid so that they may be moved as one unit and remain connected. Rig builders usually place them close to each other in an engine room. At the same time, machinery and equipment to which the engines give power are placed some distance away. Because this equipment is located many feet (metres) away from the engines, there must be a way to send, or transmit, engine power. Offshore, rig builders transmit engine power via electricity. That is, the diesel engines put out mechanical power.

The mechanical power turns electrical generators. The generators produce electricity, which heavy-duty cables carry to powerful electric motors on the equipment needing power. Drillers control the transfer of power with controls and switches from their position on the rig floor.

GENERATORS AND MOTORS

The power system on offshore rigs is usually *diesel electric*. This term means that the engines (the *prime movers*) on the rig are diesels. The diesels, in turn, drive generators to make electric power. Cables transmit the electric power to motors to drive equipment.

Figure 54.   An electric motor provides power to a mud pump.

Figure 53.   This generator, attached directly to an engine, makes electric power. *(Courtesy Sundowner Offshore Services, a Nabors Industries Company)*

In short, the diesel-electric power system consists of diesel engines, generators, and electric motors.

Rig builders attach a generator to each engine, and as an engine runs, it turns the generator (fig. 53). The generator makes electricity and powers the electric motors. Rig builders mount one or more motors on or near the device that requires power—the mud pumps, for example (fig. 54).

A diesel-electric system is very efficient. By turning electric switches, drillers can easily direct and control the power needed for every piece of equipment from a single, central control station (fig. 55).

Figure 55.   The driller controls rig power from a station on the rig floor. *(Courtesy Sedco Forex Schlumberger)*

A PRIMER OF OFFSHORE OPERATIONS

Hoisting System

Another major system on the rig is the *hoisting system*, which is a very large block-and-tackle system (fig. 56). It consists of the drawworks, or hoist; a mast or a derrick; the crown block; the traveling block; and wire rope.

Above the drill collars is drill pipe (fig. 58). As with drill collars, manufacturers make drill pipe in joints. Several joints made up together constitute the drill pipe string. The most common length of a joint of drill pipe is about 30 feet (9 metres) long.

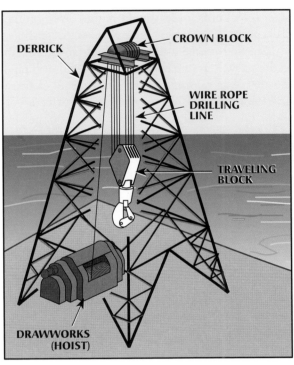

Figure 56. Crown block, wire rope drilling line, traveling block, derrick, and drawworks are the basic hoisting system parts.

Figure 57. Drill collars are heavy pipe that put weight on the bit.

THE DRILL STEM

To understand the hoisting system, let's start at the bottom of the hole, where the bit is turning. Recall that the rotating bit must have weight put on it so its teeth can cut into the rock formation being drilled. Special heavy-weight pipe called drill collars provide weight on the bit (fig. 57).

Each drill collar joint is 30 or 31 feet (9.1 or 9.4 metres) long. Several lengths, or joints, make up the string. The exact number of collars in the string depends on how much weight the collars must place on the bit: more weight, more collars; less weight, fewer collars. In any case, the weight of the drill collar string presses down on top of the bit, which crew members screwed into the bottom end of the bottommost drill collar.

Figure 58. Racked in the derrick, this drill pipe is ready to be run back into the hole.

A *stand* of drill pipe is usually three joints made up together. When the crew removes drill pipe from the hole—when they trip out of the hole—they break out (take apart) the string in stands of three joints, rather than a single joint at a time. *Breaking out* drill pipe and drill collars in stands of three joints saves time for crew members because they do not have to break out every single joint. The same thing holds true when they trip pipe and collar stands back into the hole.

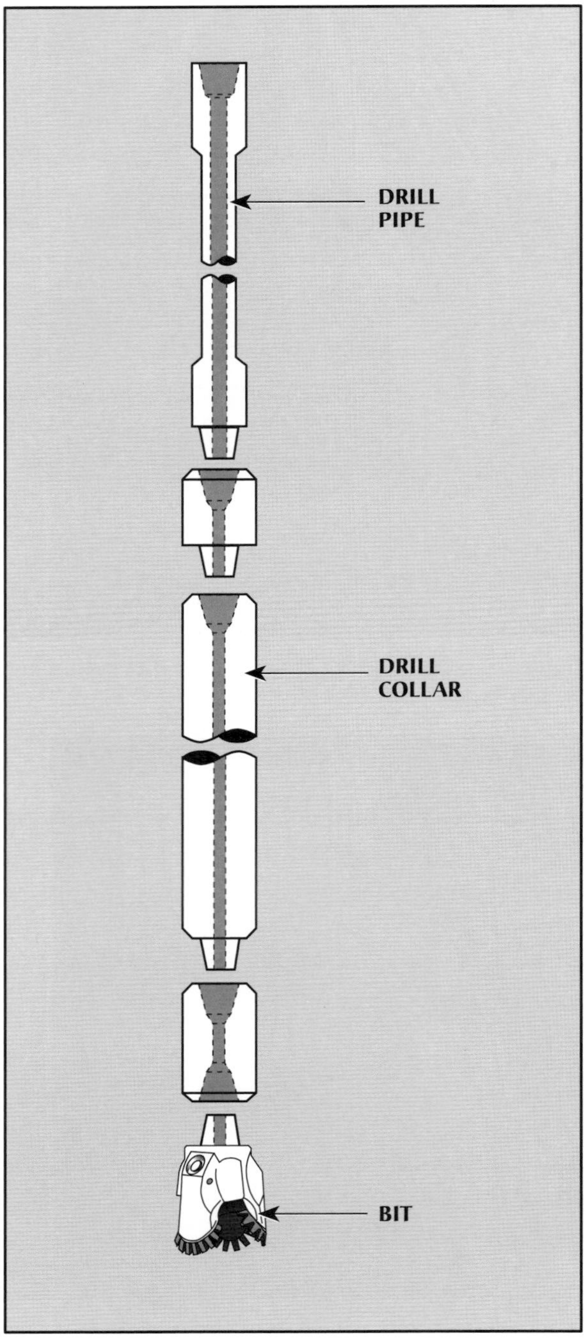

Drill pipe is not as heavy as drill collars and does not put weight on the bit. Instead, it conducts drilling mud and imparts rotary motion to the drill collars and bit. Thus, the rig has to hang, or suspend, the drill pipe above the drill collars. At the same time, it also has to exert enough upward force on the drill pipe so that none of its weight pushes down on the drill collars. The hoisting system makes this possible.

Starting at the bottom of the hole and going up to the surface are the bit, drill collars, and drill pipe (fig. 59). The hoisting system suspends all of these drill stem members in the hole; however, the driller allows only the drill collars to put weight on the bit.

HOOK AND TRAVELING BLOCK

A large, very strong *hook* supports the weight of the drill string and the top drive. The hook also hoists the drill string during a trip. Crew members attach the hook to a large set of pulleys called the *traveling block* (fig. 60).

Figure 59.   A typical drill stem assembly

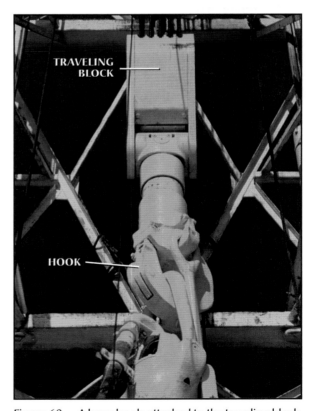

Figure 60.   A large hook, attached to the traveling block, supports the weight of the drill string. *(Courtesy Nabors Drilling, USA)*

DRILLING LINE, BLOCKS, AND DERRICK

Also very strong is special cable, or wire rope, called *drilling line*. Crew members *reeve*, (thread) drilling line through the *sheaves*, or pulleys, of the traveling block (fig. 61). (In the oil patch, sheave is pronounced *shiv*.)

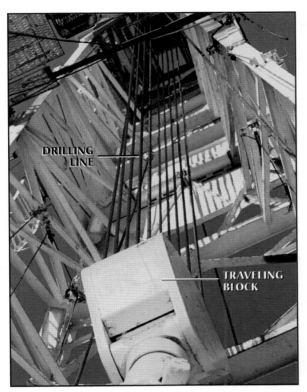

Figure 61. Wire rope drilling line reeved through the traveling block's sheaves *(Courtesy Nabors Drilling, USA)*

The drilling line passes through the traveling block sheaves and runs upward to another set of sheaves known as the *crown block*. Rig builders install the crown block near the top of the mast or derrick (fig. 62).

A *mast* or a *derrick* is the steel, open girderlike structure that makes any drilling rig so distinctive. It is tall, sometimes 200 feet (60 metres) or more, and serves as a structural tower. It supports the crown and the traveling blocks and the weight of the drill stem (fig. 63). The derrick must be tall enough to allow crew members to manipulate stands of drill pipe and drill collars, which are about 90 feet (27 metres) long, inside of it.

A mast and derrick are different, even though most people in the oil industry call a mast a derrick. Strictly speaking, a mast is a portable derrick that crew members can raise as an entire unit. A standard derrick has four legs standing at the corners of the rig's substructure. Crew members cannot raise a standard derrick as a unit. They have to build it piece by piece. So, a standard derrick is not portable. Nevertheless, this book will use the terms derrick and mast interchangeably, as most persons in the oil industry do.

Crew members reeve the drilling line through the sheaves in the traveling block and the sheaves in the crown block. They reeve the line several times between the two blocks. The drilling line is a single continuous line. However, crew members run it back and forth several times between the traveling and crown blocks. The effect is that of several

Figure 62. The crown block at the top of a mast *(Courtesy Nabors Drilling, USA)*

CROWN BLOCK

DERRICK

TRAVELING BLOCK

Figure 63.   The derrick on this drill ship supports the crown and traveling blocks and the drill stem.

Figure 64. The drilling line is reeved between the traveling block and the crown block to increase hoisting capacity. Ten lines are shown here. *(Courtesy Nabors Drilling, USA)*

lines—eight, ten, or even twelve. So, if crew members say their rig has ten lines strung, they really mean that it only has one line. However, they reeved it five times between the traveling and crown blocks, which gives the effect of ten lines. Running the drilling line several times between the blocks multiplies the hoisting capacity of the line (fig. 64).

DRAWWORKS AND DRILLING LINE

After reeving the drilling line between the crown and traveling blocks the required number of times, crew members then take another step. They run the end of the drilling line downward from the crown block and onto a large spool that turns inside the hoist, or

Figure 65. The drawworks is a large hoist. *(Courtesy Nabors Drilling, USA)*

drawworks (fig. 65). The drilling line wraps around the drawworks spool, much as thread wraps around a wooden spool (fig. 66).

When the driller lets off drilling line from the spool, it unwinds to allow the traveling block to lower. Conversely, when the driller takes in line on the spool, the line raises the traveling block. Also, the driller adjusts the weight on the bit by letting out or taking in drilling line a small amount. Taking in and letting out large amounts of drilling line allows crew members to *trip* the entire drill stem in and out of the hole. They have to trip the drill stem for various reasons. A common one is to change out a dull bit.

In summary, the hoisting system does two things. For one, it suspends the drill stem in the hole to apply the desired weight on the bit. For another, it allows the driller to lower and raise the drill stem into and out of the hole.

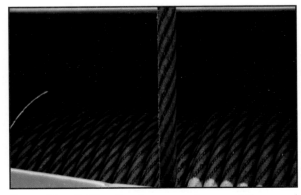

Figure 66. Drilling line is wrapped around the spool (drum) of the drawworks. *(Courtesy Nabors Drilling, USA)*

Drilling Personnel

The people in offshore drilling work for a variety of companies that engage in activities of many kinds. The total number of people who may be present on the rig depends on the size of the unit and what work it is doing. Typically, a large mobile offshore rig may have 100 or more people on board during peak activity. A small jackup working close to shore may have fewer than 20 persons. The usual complement falls somewhere between the two extremes.

OPERATOR

The *operator*, or operating company, buys or leases the right to explore for, drill for, and produce hydrocarbons. The expenses of drilling and producing offshore wells are so enormous that often more than one company is involved in a venture.

For example, four companies—Conoco, Union Oil, Shell Oil, and Superior Oil (now absorbed by another oil company)—got together to develop one of the first drill ships. Named *CUSS I*, an acronym of the company names, the ship did the first exploration drilling off the California coast. Since exploring the continental shelf in this area was a very expensive proposition, the four joined and called themselves the CUSS group.

Company Man

The operator almost always has a representative on the rig at all times during the drilling operation. This person may be called the *company man*, *rig manager*, or *rig superintendent*. The company man oversees the entire drilling operation and makes company decisions and policies, relaying them to the drilling contractor's top man—the toolpusher. In general, the company man sees to it that the contractor drills the well to the specifications set by the company.

DRILLING CONTRACTOR

Even though an oil company or companies actually own the well to be drilled, they hire many of the services needed to drill it. For instance, a drilling contractor usually owns the rig that does the drilling. A *drilling contractor* is a firm that specializes in drilling wells, sometimes both on land and offshore.

Drilling Crew

The drilling contractor hires the people who make up the drilling crew. Generally, two crews work 12-hour shifts called *tours* (pronounced *towers*). Often, these two crews work for seven days and are off seven days, at which time another two crews come on duty to replace them.

You can find a lot of variation on the seven-on, seven-off scheme, however. For example, if a U.S. drilling contractor has a rig working in a remote area such as the South China Sea, the U.S. citizens working on the unit may rotate on a 28-day basis. This arrangement allows the crews time to travel to and from their homes in the United States.

Toolpusher. As for who's who on a typical drilling crew, the contractor's top hand on the rig is the *toolpusher.* The toolpusher supervises the drilling crew and coordinates oil company and drilling contractor affairs on the rig. Sometimes the toolpusher may have an assistant, a night toolpusher or assistant toolpusher, who is on board to relieve and help as required.

Driller and Assistant Driller. The *driller*, under the direct supervision of the toolpusher, is responsible for the actual operation of the machinery on the rig (fig. 67). The driller is also responsible for the performance of the drilling crew. In many cases, an assistant driller works directly under the driller, especially in countries other than the United States. The assistant driller relieves the driller when necessary and is in training to become a driller.

Figure 67.   A driller works at his station on the rig floor. *(Courtesy Sedco Forex Schlumberger)*

Derrickhand. The *derrickhand* handles the upper end of the drill stem when it is being tripped in or out of the hole. The derrickhand does this from a small working platform up in the derrick called the *monkeyboard* (fig. 68) and also does routine monitoring of the drilling mud when the bit is on bottom and drilling.

Pit Watcher. Some offshore rigs designate a person as a *pit watcher*. The pit watcher's jobs include assisting the derrickhand in keeping tabs on the drilling mud and the circulating equipment.

Rotary Helpers. *Rotary helpers*, also called *floorhands* or *roughnecks* work on the rig floor and handle the lower end of the drill stem stand when it is being tripped. Usually, the contractor employs three rotary helpers, but the number can vary (fig. 69). They are also responsible for keeping the rig floor clean and doing routine maintenance on the drilling equipment.

Rig Mechanic and Electrician. Rig mechanics and electricians maintain and repair the engines and mechanical and electrical equipment that abound on any offshore rig.

Roustabout Foreman and Roustabouts. The *roustabout foreman*, or head roustabout, is the supervisor of the roustabout crew. A roustabout crew's duties

Figure 68. Standing on a small platform high in the derrick (the monkeyboard), a derrickhand guides the top of a stand of drill pipe. Note the safety harness to prevent a fall.

Figure 69. Three rotary helpers make up a joint of drill pipe. *(Courtesy National-Oilwell)*

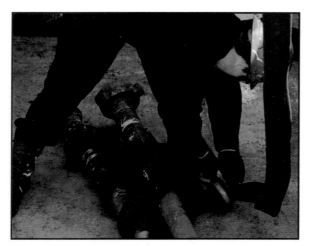

Figure 70. A roustabout attaches a lifting sling to a crane's hoist line. *(Courtesy Sundowner Offshore Services, a Nabors Industries Company)*

consist of routine cleaning and maintenance of the entire rig. The crew also assists in the loading and unloading of boats bringing supplies to and from the rig (fig. 70). Usually, the roustabout foreman is also the *crane operator* (fig. 71). The rig uses large

cranes to hoist and move equipment on the rig and to lift and load supplies to and from the boats.

Ballast-Control Specialist. The drilling contractor employs *ballast-control specialists* on semisubmersibles to control the stability of the unit while it is on location and drilling. Ballast-control specialists carefully note the placement of heavy equipment and supplies on the deck of the vessel. They also control the quantity and placement of water used as ballast. A ballast-control specialist's main job is to maintain the rig in a stable condition at all times (fig. 72).

OTHER COMPANIES

Operators also hire helicopter and boat companies to transport personnel and supplies to and from the offshore location (fig. 73). In addition, they employ personnel from service and supply companies to provide and maintain special equipment and techniques needed to drill the well. An example of a service company is a mud company, the company that supplies and maintains the drilling mud.

Figure 71. Usually, the roustabout foreman also operates the large cranes on the rig. *(Courtesy Sundowner Offshore Services, a Nabors Industries Company)*

A PRIMER OF OFFSHORE OPERATIONS

Figure 72. A ballast-control specialist stands at the ballast-control panel of a semisubmersible. The layout gives him a visual indication of the location and amount of ballast in the rig's pontoons.

Figure 73.   Helicopter and boat companies transport personnel and supplies. *Top,* personnel transfer from the rig to the boat; *bottom,* a helicopter lands on the rig's helipad.

The operator may hire a diving contractor if the drilling operation requires divers to install or inspect special subsea equipment on the well. The operator or contractor will usually hire a catering company to supply and prepare food for personnel who work and live on the offshore rig.

On floating rigs and on bottom-supported units being moved, a qualified ship's captain, who is responsible for handling the vessel when it is under way, is on board. Under his or her supervision are ship's officers and able-bodied seamen.

EMERGENCY RESPONSE

The drilling contractor and operating company provide emergency instructions for all personnel on board the facility. All hands thoroughly read the instructions to find out precisely where to go and what to do when they get there. While emergencies are rare, they do occur, so everyone must be prepared. Thus, the rig owner posts a *station bill* (fig. 74). The station bill covers various types of emergencies, such as fire, blowout, or rig abandonment. It instructs each person in

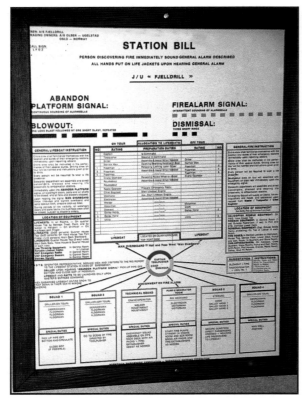

Figure 74.   This station bill tells rig personnel how to recognize emergency alarms, and gives duty and station assignments.

emergency procedures, and explains the alarms that will sound for each type of emergency.

Emergency duties and duty stations vary depending on the type of emergency. For example, in case of fire, instructions may direct the crane operator and roustabouts to assemble at a specific spot on the rig. Once there, they must put on the proper protective gear and be ready to operate fire extinguishing equipment. On the other hand, if the person in charge sounds the abandon rig alarm, everyone reports to an assigned life boat, which offshore personnel call an *escape capsule* (fig. 75). To ensure readiness, crew members perform regularly scheduled (as well as unscheduled) emergency drills to be ready for the real thing.

Figure 75.   Underway during a rig abandonment drill, personnel steer an escape capsule away from the rig.

OFFICE PERSONNEL

An important group of people involved in offshore drilling work mostly on shore, in the relatively ordinary world of an office building. Office work often pales when compared to the drama and excitement of the work on an offshore rig. Yet without the efforts of dedicated personnel who rarely, if ever, set foot on an offshore facility, the industry probably could not function.

Ranging from company executives to clerical staff, office personnel provide essential support to those in the field. The international nature of offshore drilling increases the problems faced by the office team. For example, since all those far-flung rigs must be staffed by qualified personnel, the hiring and training of these people is a large concern of any company.

Logistics—that aspect of offshore operations that deals with obtaining, transporting, and maintaining rigs, crews, and equipment—is also a major concern of a company's office staff. The logistical problems facing a company whose rigs are scattered all over the world are phenomenal. The company has to obtain passports, visas, certificates, and work permits for its rig crews. Further, it has to plan and execute the transporting of personnel and equipment to the rigs. Moreover, the company must feed and house its rig crews, provide for their medical needs, and keep accurate records. The list is long. Thus, without a competent team of office personnel working closely with those in the field, a company would drill very few offshore wells.

Summary

The primary job of any drilling unit is to put a drill bit into the ground and rotate it. To accomplish the task, the rig must circulate drilling fluid to clean cuttings from the bottom of the hole and to carry the cuttings up the hole for disposal. In addition to a circulating system, the rig also needs a rotating system to turn the drill stem and bit. Rigs either use top drives (power swivels), a rotary table system, or bottomhole mud motors to rotate the drill stem and bit.

Several diesel engines provide rig power. The engines turn generators to generate electric power. Cables from the generators carry electricity to motors mounted on or near the equipment to be powered.

The hoisting system consists of the drawworks, crown block, traveling block, and drilling line. The hoisting system suspends the drill stem and bit in the hole and makes it possible to trip the drill stem and bit in and out of the hole.

The operator is an oil company that buys or leases the right to explore for, drill for, and produce hydrocarbons. The operator usually hires a drilling contractor—a firm whose specialty is drilling wells and one that owns drilling rigs—to drill a well. Operators also hire many service and supply companies to provide additional equipment and services they need to drill a well.

5

EXPLORATION DRILLING

Once the operator has acquired the rights to drill and has determined precisely where to locate the rig on the drill site, the exploration drilling phase of offshore operations begins. The operator first selects a rig. With the rig selected, the next step is to move it onto the site and actually drill the well.

Whether the operator hires a bottom-supported unit or a floater, the primary idea is to get the bit turning on bottom with weight on it while circulating drilling mud. Enough differences exist between the two types of rigs, however, that we will discuss them separately.

Throughout this discussion on drilling from bottom-supported units and floaters, you will read about typical operations. Keep in mind that specific drilling techniques vary from area to area; what works in one part of the world may not be suitable elsewhere. Basic techniques are similar the world over, however, and they are emphasized.

Selecting a Rig

Generally, companies can drill an offshore exploratory well from one of two types of rig. One is a bottom-supported unit, such as a jackup or a submersible. The other is a floating unit, such as a drill ship or a semisubmersible. The operator's choice depends on many factors, including water depth, load capacities, the weather and other environmental conditions, and rig availability.

WATER DEPTH

Water depth plays an important role in rig selection. For example, in deep water, over 350 feet (100 metres) or so, the operator usually selects a floating unit. The water is too deep for most bottom-supported units.

LOAD CAPACITIES

A drill ship can carry much greater deck weight than a semi. A semi's vertical columns do not displace much added volume of water as the rig is loaded. So, when crew members add more weight, the semi submerges more deeply into the water. A great deal of weight on the drilling deck of a semi affects its ability to remain upright, even in calm seas.

A drill ship, on the other hand, displaces a lot of water, with only a small additional push down into the water when loaded. Thus, it is not as likely to capsize when subjected to heavy loads. It can carry most of the equipment needed to drill a well and does not need to be resupplied as often as a semi does. Since it is less dependent on supply boats, it can operate in remote waters for long periods.

WEATHER AND OTHER ENVIRONMENTAL CONDITIONS

Weather conditions also play a large role in rig selection. For example, in relatively shallow water where a jackup or submersible would be suitable under calm conditions, a semisubmersible might be a better choice if the weather is adverse.

Most rigs working in the North Sea are semis because of the rough seas that prevail in that part of the world. Even though water depths often do not exceed a jackup's capabilities, rough seas make it difficult to get a jackup on location. Also, at times the sea runs so high that jackups cannot withstand the tremendous forces generated by the waves.

Because a drill ship floats on the surface, wave motion affects it more than a semi. Thus, for drilling in calm, remote waters, the operator would more likely select a drill ship instead of a semi.

In offshore arctic drilling, special considerations for pack ice change the rules of the game. For most of the year, the seas around and north of the Arctic Circle are frozen. The ice does not form as a continuous sheet. Instead, it breaks up into massive chunks of ice that move with wind and currents. These heavy clumps of ice can destroy all but the strongest structures. To withstand the forces of moving ice, operators usually employ specially designed submersibles, like the mobile arctic caisson (MAC).

In shallow arctic waters, as in the Beaufort Sea off Alaska's North Slope, operators once constructed *artificial gravel islands* (fig. 76). Construction contractors made these islands in several steps. First, they built an ice road to the location. They formed a thick, high-load-bearing ice road by pumping large amounts of sea water onto the existing pack ice. Once on the pack ice, the sea water froze to form the road. Second, the contractors cut a large hole in the ice. Third, they hauled tons (tonnes) of gravel out to the site by trucking it over the ice roads. Fourth, they filled the hole with gravel. Fifth, the contractors placed large gravel-filled sacks around the perimeter of the island to divert flowing ice around it. Finally, the drilling contractor trucked out a winterized land rig to the island.

Building such islands took enormous amounts of gravel, which the contractor had to collect on land. Not only did it cost operators a lot of money to pay for the gravel, but they also had to spend more money to repair the environment where they removed the gravel. Because of the large costs of the artificial islands, operators are using mobile arctic rigs instead.

AVAILABILITY OF RIGS

An offshore operator may select a rig simply because it is available when needed. Thus, in areas where water depth and weather are not big factors, you can find any of the mobile offshore drilling units at work. For example, in the Gulf of Mexico, relatively calm seas are the rule and in many active areas the water is not too deep. (There is also deep water work as well.) You may therefore see jackups, submersibles, drill ships, and semisubmersibles working in this region.

Figure 76.   An artificial gravel island and its drilling rig loom out of the fog in Alaska's Beaufort Sea. The sea is completely frozen and thus shows up as white behind the island. *(Courtesy Shell Oil Co.)*

Drilling from Bottom-Supported Units

To move a bottom-supported unit onto the drill site, rig owners first put the unit into a floating mode. They then (in most cases) use tug boats or similar vessels to tow it to the site. A few bottom-supported units are self-propelled. And if drilling contractors have to move a unit a long distance, they may load it on a special ship (see fig. 22).

Once on location, crew members put the unit into the drilling mode. They flood the bottles or jack down the legs to put the unit in firm contact with the seafloor.

CONDUCTOR CASING

The next step involves the use of special large-diameter steel pipe called *casing*. Casing is essential to the drilling of all wells. The well owner uses it to line the raw, open hole made by the bit in the formations. It keeps the hole from caving in and, when crew members cement it in place, casing keeps fluids in one formation from migrating to another. Thus, oil, for example, cannot flow into a freshwater zone and contaminate it.

Bear in mind that contractors drill holes in steps. The bit drills the hole to a certain depth, and crew members remove the bit and drill stem from the hole. Casing is then put into the hole to line it, and, in most cases, cement is pumped in to bond the casing firmly to the walls of the hole. The rig drills more hole to a certain depth and the crew cements another casing string into the hole, and drilling begins again. Drilling a portion of the hole and lining it with casing and cement continues until the rig reaches total depth.

Lining the hole with casing and cement serves several purposes. For one thing, casing and cement keep the hole from caving in after it is drilled. For another, any liquids or gases that may be confined in the formations are kept out. Finally, casing provides a conduit from the seafloor to the drilling unit on the surface of the water.

Casing, like drill pipe and drill collars, is manufactured in *joints*, and each joint ranges in length from 16 to 48 feet (4.88 to 14.63 metres) (fig. 77). Casing joints range in diameter from 4.5 to 48 inches or more (114.3 to 1,219.2 millimetres or more). Crew members screw each casing joint together or, in some cases, weld it together as they lower it into the hole. Several joints of casing, when joined, constitute a casing string.

To *spud in*, or start the hole, from a bottom-supported unit, operators can use a couple of different approaches. The one selected depends to a large extent on the composition of the seabed.

If the bed is soft, the first step won't even involve a drill bit. Instead, crew members jet or hammer the casing into the seafloor. This casing is large in diameter—perhaps 26 inches (660.4 millimetres) or larger—and is called *conductor casing*, or *conductor pipe*. It is called conductor pipe because it forms the top part of the well and conducts drilling mud back up to the surface.

Figure 77.   Two joints of casing lie on a ramp, ready to be picked up and run into the hole.

To jet conductor casing into place, the rig pumps drilling fluid, usually seawater, down inside the casing as crew members lower it onto the seafloor. The high-pressure stream of water erodes the soft sea bottom and creates a hole for the casing for the crew to lower it into.

To hammer conductor casing into place, crew members rig up a large, powerful pile driver in the derrick of the drilling unit. The pile driver literally hammers the casing into the seabed. Usually, this first string of casing isn't very long—perhaps a few hundred feet (metres) at most. So, jetting or hammering conductor casing in place works very well in soft, unconsolidated ooze or mud.

Where harder material (like coral or hard clay) makes up the seafloor, crew members have to drill a hole first and run and cement the conductor casing into the drilled hole. In this case, they lower a large-diameter drill bit on drill pipe into the water until it comes to rest on the seafloor. The driller starts rotating the bit, breaks circulation (starts the mud pump to circulate drilling fluid), and drills the hole. The cuttings simply come back up the hole and fall onto the seafloor. When the hole reaches a predetermined depth, crew members remove the bit and drill stem, run conductor casing into the drilled hole, and cement it in place.

The casing's top extends upward to a point above the water line and below the drill floor of the drilling unit (fig. 78). This first string of casing extends the hole from the seafloor, up through the water to a point in the air just below the drilling deck of the rig. The conductor casing also provides stability to the part of the hole that is near the seafloor.

Figure 78.   A worker stands on the conductor casing that extends upward from the hole in the seafloor.

A PRIMER OF OFFSHORE OPERATIONS

DRILLING OPERATIONS

Once crew members jet or hammer the conductor casing in place, they lower a bit small enough in diameter to fit inside the casing. They then drill out any soil and ooze that may remain in the casing. Drilling continues past the bottom of the conductor pipe to some predetermined depth, usually a few hundred feet (metres).

If the crew lowered and cemented the conductor casing into a drilled hole, there is no soil inside the casing, so they lower the bit inside the casing to bottom, and drill the hole deeper. As is the case with jetted or hammered casing, drilling continues to a predetermined depth of a few hundred feet (metres). In either case, drilling comes to a temporary halt at this point.

To review a moment, think of a string of large-diameter steel pipe—conductor casing—sticking up out of the water under the rig. It runs down through the water and into the seafloor to some depth, usually dozens of feet (metres), if not hundreds.

Crew members lowered a bit through the inside of the casing and drilled out a raw, open hole to some depth below the casing again, probably only a few hundred feet (metres).

They removed all of the drill stem and the bit from the hole, and drilling has come to a temporary halt (fig. 79). Yet the formation that the oil operator hopes will contain oil or gas lies perhaps thousands of feet (metres) below.

Why has drilling stopped? Because the crew has to run more casing into the hole. A well penetrates many types of rock formations, or zones, on its way to the formation that may contain hydrocarbons. Some of the formations may contain fluids like oil, gas, and water under very high pressure; others may have a tendency to *slough* and cave into the hole. Because formations are different, the crew cannot drill a well in a single step from surface to total depth. They must run casing into the hole to protect upper zones from the drilling muds needed to drill lower zones.

Drilling Mud

Crew members carefully formulate and control drilling muds to drill various formations. For instance, the crew can drill formations that contain gas or water under high pressure with drilling mud that has high weight, or density. The more a mud weighs—the denser it is—the more pressure it exerts

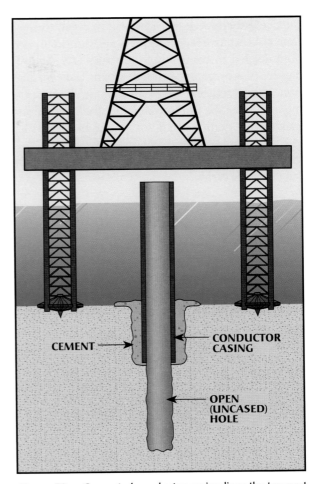

Figure 79. Cemented conductor casing lines the top part of the hole. More hole has been drilled out below the casing.

on formations exposed to the wellbore. Heavy mud exerts pressure in the hole, and this pressure keeps fluids (gas, water, or even oil) from leaving a formation and entering the hole. If the crew does not hold formation fluid pressure in check, formation fluids can enter the well.

When formation fluids enter the well, crew members say that the well has kicked. If they do not take immediate steps to control the kick, more formation fluids can enter the well and the well can blow out. A blowout is the uncontrolled flow of formation fluid to the surface or into an underground formation. If fluids blow out to the surface, they can severely damage or destroy the drilling unit.

Unfortunately, mud that works for deep formations may not work for formations at shallower depths. For example, the hole may need heavy mud to keep high formation pressure contained at some relatively shallow depth. However, a weak formation may occur deeper in the hole, and the heavy mud needed to control high pressure at a shallow depth could break open (fracture) this weak formation. If the fracture were severe enough, all the drilling mud could be lost into the fractured formation, making it very difficult, if not impossible, for drilling to continue. The oil operator formulates the mud to drill the upper formations, then stops drilling and runs casing.

ADDITIONAL CASING STRINGS

Once the crew cements casing in the hole, the casing and cement protect the upper zones from further exposure to the mud needed to drill the lower zones. In deep wells, the well may require two, three, or even four strings of casing (in addition to the conductor casing) (fig. 80).

Crew members have to run these multiple strings of casing one within the other. What is more, there must be room between the walls of each string of casing and the wellbore for cement. So, the diameter of each string the crew runs gets progressively smaller. For example, with 30-inch (762-millimetre) conductor casing, each subsequent casing string the crew runs is smaller in diameter. The final string may thus be only 6 inches (152.4 millimetres) in diameter.

In summary, the drilling crew does not drill a well from the surface to the reservoir in a single stage. Instead, the crew sets conductor pipe and starts drilling the hole. Then they run and cement another string of casing (the *surface casing*) and drill more hole. Next, they run and cement still another string of casing (the *intermediate casing*) and drill still more hole. Finally, the crew runs and cements one last string of casing (the *production casing*) at the well's total depth.

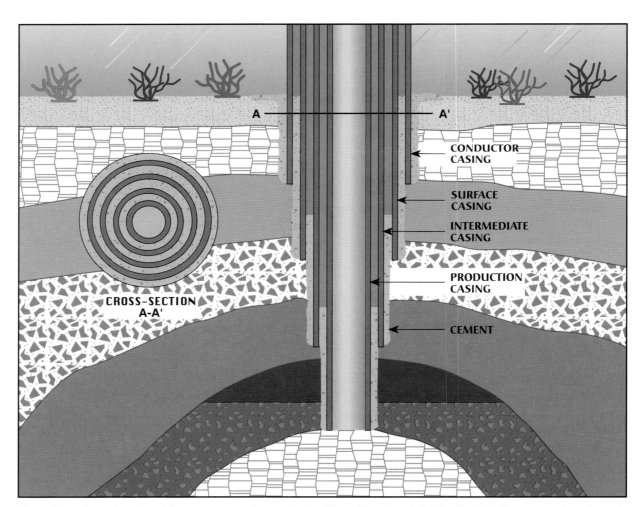

Figure 80.   Several casing strings are run and cemented as the well reaches total depth. Note the names given to each string.

A PRIMER OF OFFSHORE OPERATIONS

BLOWOUT PREVENTION

Besides lining the walls of the hole and isolating formations, casing also serves as a mounting place for the *blowout preventers (BOP)*. Crew members usually mount several blowout preventers on the casing, one preventer on top of the other. Since they stack the preventers, they call the whole assembly of preventers the *blowout preventer stack (BOP stack)*.

If the crew does not control formation fluid pressure, the pressure can enter the drilled hole and cause a blowout. Usually, crew members weight the drilling mud so that its density creates enough downhole pressure to confine fluids in the formation. Occasionally, however, the borehole encounters unexpected pressure and the well *kicks*: gas, water, or oil enters the borehole. Fortunately, when a kick is about to occur, it gives certain warning signs on the rig. An alert drilling crew can detect kick warning signs and take proper steps to prevent the well from blowing out.

Part of the procedure for controlling a kick involves the use of large, quick-closing valves. These valves are the blowout preventers. On bottom-supported units, crew members install them beneath the drilling floor and on top of the casing that extends up from the wellbore in the seafloor (fig. 81).

When the crew recognizes a kick, the usual steps are to shut in (close) the BOPs and remove the intruded formation fluids from the hole. They then increase the mud's density to prevent further fluid entry, and resume normal drilling procedures.

BOPs are very important to the drilling of all wells, especially those offshore, where blowouts can pollute the sea, damage extremely expensive equipment, and, worst of all, threaten human life. That is why drilling crews receive thorough training in kick detection and control. Indeed, most countries require that crews be certified in well control.

Figure 81.   These blowout preventers are mounted on the casing. *(Courtesy Sundowner Offshore Services, a Nabors Industries Company)*

Drilling from Floating Units

Drilling from floaters (drill ships and semi-submersibles) is similar to drilling from bottom-supported units (submersibles and jackups) with a couple of major exceptions. Floaters move with the wind and current action of the sea; they heave, pitch, roll, yaw, surge, and sway (fig. 82). These movements affect the ability of a floater to drill efficiently. As a result, the rig owner must compensate for them to successfully drill a well.

Heave, the tendency of a floater to move up and down, is one of the more important movements to compensate for. The driller must keep the weight on the drill bit as constant as possible, at the required value, for the bit to work at its best. Heave tends to pull weight off the bit as the floating rig rides to the crest of a wave and puts weight back on the bit as the rig wallows into a trough between waves. How the rig compensates for heave is covered later.

Another major difference between drilling from floaters and drilling from bottom-supported units is that crew members place a floater's blowout preventer stack on the seafloor. Remember that naval architects designed floaters to drill mainly in waters too deep for bottom-supported units. Putting the BOP stack

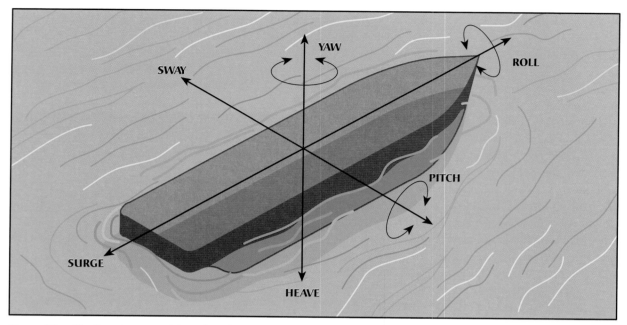

Figure 82.   Floating rigs are subjected to six motions caused by the wind, current, and wave action of the sea.

on the seafloor does not permit the use of a long string of casing from the seafloor up to the rig. If not supported by a surrounding wellbore and a sheath of cement, a long string of casing will buckle and fail. It is simply too heavy to support itself if it exceeds a certain length.

Since floaters usually drill in water depths that exceed the ability of casing to support itself, engineers developed a subsea BOP system. A floater with a *subsea BOP system* uses special pipe called *marine riser pipe*. The marine riser system serves as a conduit from the subsea BOP equipment to the drilling floor of the floater. Put another way, the marine riser provides a connection between the well on the seafloor and the drilling rig.

A typical drilling operation on a floater follows. Again, remember that techniques can vary from drill site to drill site.

PREPARING TO DRILL

Once the rig owner positions a floater over the drill site, either with anchors or by dynamic positioning, operations can begin. First, crew members attach a heavy steel framework to the bottom end of a string of drill pipe. They then lower the framework to the seafloor. The framework is a *temporary guide base*, or *drilling template*. It is a heavy steel device that is hexagonal with a circular opening in the center (fig. 83).

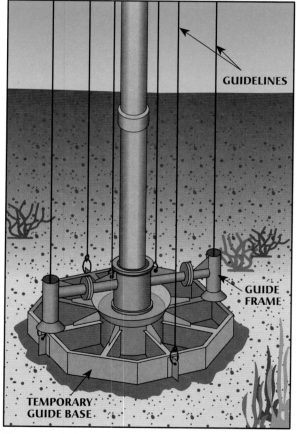

Figure 83.   The temporary guide base is landed on the seafloor by drill pipe. Note the four guidelines running up to the rig.

In spite of its name, nothing is really temporary about it. Once the crew places it on the seafloor, it will likely remain there during the life of the well. The temporary guide base has an opening in the center. Attached on its perimeter are four cables, called *guidelines*.

The drill pipe on which crew members lower the temporary guide base fits into the center opening. When the drill pipe lands the guide base onto the ocean bottom, serrated legs on the bottom of the base penetrate the seafloor and keep the base stationary.

Crew members then remove the drill pipe from the temporary guide base and pull the pipe back up to the floating rig. Its removal leaves the guide base on bottom with four guidelines running back up to the rig.

DRILLING OPERATIONS

With the guide base in position on the seafloor, the next order of business is drilling a hole for the first string of casing. In floating operations, the first string of casing is commonly called the foundation pile. Foundation-pile casing is the same thing as conductor casing. In floating operations, however, crew members use the term *foundation pile* to designate the first string of casing run into the hole. Because the casing that makes up the foundation pile is usually 30 or 36 inches (762 or 914.4 millimetres) in diameter, crew members use a bit and hole opener to drill the hole.

The hole must be larger in diameter than the casing so that crew members can run casing into it. Since the bit may be only 17½ inches (about 444.5 millimetres) in diameter, crew members make up a *hole opener* in the drill stem above the bit (fig. 84).

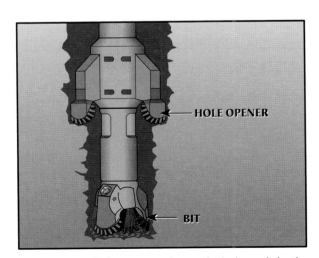

Figure 84.   A hole opener enlarges the hole made by the regular bit.

The hole opener enlarges the hole made by the bit to the required diameter, 30 or 36 inches as the case may be.

Crew members make up the bit and the hole opener on drill collars and drill pipe as usual. In addition, however, the crew installs a *guide frame* on a drill collar joint near the bottom (fig. 85). The guide frame has two, or sometimes four, arms through which crew members thread guidelines.

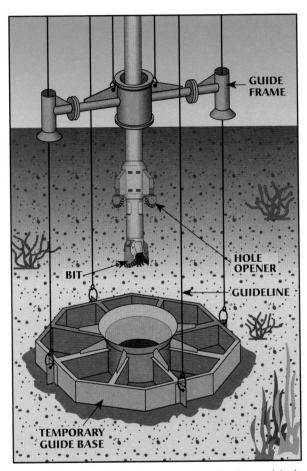

Figure 85.   The guide frame centers the bit and hole opener into the opening in the temporary guide base.

Guidelines are relatively small-diameter wire ropes. Crew members attach them to the guide frame on the seafloor and to a special fitting near the rig floor on the surface. The crew lowers the bit and drill stem into the water. The guidelines keep the guide frame, and thus the drill stem and bit, in line with the center opening in the temporary guide base.

When the bit reaches the seafloor through the opening in the temporary guide base, the driller begins circulating and rotating, and the bit and hole opener drill the hole for the foundation pile. As the bit drills, the guide frame comes to rest on the guide base. It remains in this position until the crew pulls the bit and drill stem out of the hole. At this time, the guide frame comes up with the drill stem.

At this point, crew members have drilled the hole for the foundation pile. Resting on the seafloor is the temporary guide base with its four guidelines running up to the rig. The hole for the foundation pile is usually about 100 feet (30 metres) deep, but this depth can vary, depending on the softness or hardness of soil near the surface of the seafloor.

RUNNING FOUNDATION-PILE CASING

When the crew runs the foundation-pile casing, they lower it through the rig's moon pool, into the water, and down the hole, one joint at a time. They can weld or screw the joints together, or use special threadless couplings that allow them to make up the casing joints by pressing them together. The couplings go by the name of *Squnch Joints*™. All the crew does to connect the casing is to allow one joint of casing to "squnch down" (press down under its own weight) on top of another one (fig. 86).

The crew attaches a guide frame to one of the bottom joints of foundation-pile casing. The arms of the guide frame ride in the guidelines and guide the casing into the center opening in the temporary guide base and thus into the borehole. The guide frame arms break off as the casing enters the hole.

Crew members attach a special assembly, the *foundation-pile housing*, to the top of the topmost joint of casing before they lower the casing into the water. This housing supports the foundation pile when it lands in the temporary guide base. It also provides a landing base for the next casing string.

The crew installs the permanent guide structure around the foundation-pile housing (fig. 87). They land the permanent guide structure and housing in the temporary guide base. The guide structure and housing provide a structural base for additional casing strings and other equipment, such as the blowout preventer stack, that crew members will run later.

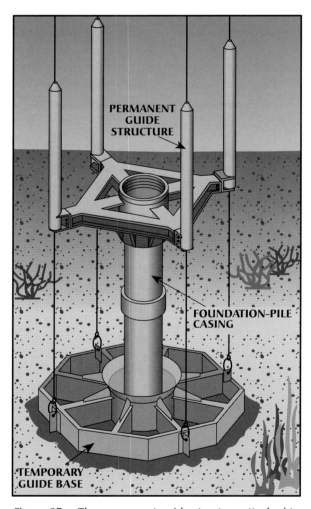

Figure 87. The permanent guide structure attached to the top of the foundation-pile housing is lowered on the guidelines. The foundation-pile casing enters the drilled hole, and the permanent guide structure seats on the temporary guide base.

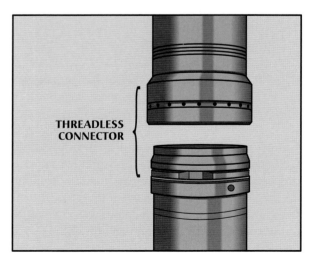

Figure 86. A Squnch Joint™, a threadless connector used to make up large-diameter joints of casing

A PRIMER OF OFFSHORE OPERATIONS

CEMENTING THE FOUNDATION PILE

Once crew members run the foundation pile into the hole and land the permanent guide structure in the temporary guide base, they cement the foundation pile in the hole.

To cement the foundation pile, the crew lowers drill pipe inside the foundation pile to a point near the bottom, where the pipe is connected to a fluid-tight seal. Cement is then pumped down the inside of the drill pipe. The cement goes out the bottom of the foundation pile and back up the annular space between the wall of the hole and the outside of the casing.

Crew members pump cement until it starts to spill out onto the seafloor beneath the temporary guide base. They then stop pumping and retrieve the drill pipe.

DRILLING MORE HOLE

After the cement sets, or hardens, the hole for the second string of casing is ready to be drilled. The second string is called conductor casing. To drill the hole for the conductor casing, crew members lower a bit into the water to the bottom of the foundation pile. The bit's diameter is a few inches (millimetres) smaller than the inside diameter of the foundation pile.

A guide frame, attached to a drill collar, guides the bit into the hole. Once the crew drills the hole to the required depth, they pull bit and drill stem, and run and cement conductor casing.

The next step is to prepare the *subsea blowout preventer* stack. Crew members pressure test it to ensure that it is operating properly. They then lower it down to the permanent guide structure (fig. 88).

MARINE RISER AND BOP STACK

The BOP stack that a floating drilling unit uses is different from that on a jackup. Typically, the subsea BOP stack has a larger bore diameter, since all subsequent casing strings must pass through it. Crew members usually assemble it in two main sections, each surrounded by a guide frame, which will slide along the guidelines to the wellhead.

The main section is the BOP stack itself. This section contains a remote-controlled hydraulic clamp, called a *wellhead connector*, and the ram preventers. Crew members can use the ram preventers to close the wellbore off around the drill

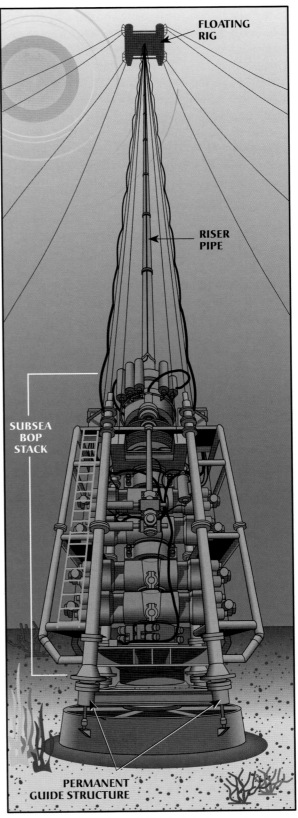

Figure 88.   The subsea BOP stack is lowered onto the permanent guide structure.

string or to cut the drill string and close in the well. The second section is called the *lower marine riser package* (*LMRP*) and contains a similar wellhead connector and an annular blowout preventer. Crew members use the annular BOP to close in the well around the drill string or, if necessary, on open hole. Open hole is a wellbore with no drill pipe in it.

After the crew assembles the BOP and the LMRP, they lower them to the seafloor on joints of marine riser pipe. Once they set the BOP stack on top of the subsea wellhead, they lock on the hydraulic wellhead connector. They then perform a pressure test. The test ensures that the sealing areas in the wellhead and the BOP work properly.

The marine riser pipe conducts drilling fluid from the well to the drilling rig. Also, since it is connected to the well, the marine riser guides all additional drill strings, casing strings, and downhole tools to the well without a guide frame.

The marine riser system consists of several parts. From bottom to top, these parts are a *flexible joint*, a *marine riser connector*, individual *riser joints*, *choke* and *kill lines*, and a *telescopic joint* (fig. 89).

The riser connector provides a way to attach the first joint of riser pipe to the flexible joint. Below the flexible joint are devices that connect the riser system to the BOP stack (see figs. 88 and 89). With the riser system installed onto the BOP stack, the flexible, or ball, joint allows the riser to move a few degrees around the connection to the BOPs. Wind and wave forces, remember, cause the floating vessel to move around its central axis. The flexible joint compensates for this circular movement.

The individual riser joints feature quick-coupling devices. Quick couplings allow the crew to make up and break out the riser joints quickly. Manufacturers attach two small pipes, the choke and kill lines, to the outside of each riser joint. The choke and kill lines run to connections on the BOP stack. They allow crew members to pump mud into the well and for fluids to exit the well when the BOPs are closed.

The telescopic joint, often called the *slip joint*, compensates for heave—the up-and-down motion of the floating rig. Essentially, it is a special tube, or barrel, that slips inside the riser pipe. The inner barrel connects to the rig and thus moves up and down with it. The outer barrel, since it is part of the riser pipe, remains stationary with respect to the seafloor.

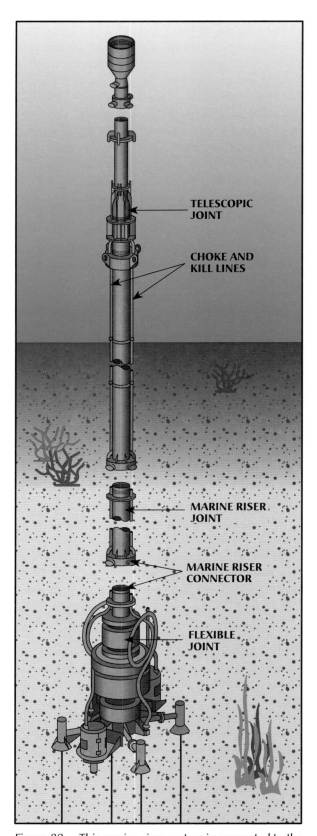

Figure 89.   This marine riser system is connected to the top of the subsea BOP stack.

MARINE RISER TENSIONING

Mounted around the outer barrel of the telescopic joint is a bracket to which crew members attach steel cables, or *wire rope*. Crew members reeve these cables through sheaves mounted on the side of the rig's moon pool. Each cable goes through a sheave and connects to a *marine riser tensioner* (fig. 90).

The riser pipe must have upward pull, or tension, applied to it to prevent it from buckling under its own weight. The riser tensioner provides this tension and allows crew members to adjust the tension. Crew members have to adjust tension on the riser pipe. The deeper the water is, the more joints of riser pipe the rig will need to reach the subsea BOP stack. The more joints of pipe used, the more the riser system weighs. And the more the system weighs, the more tension required to keep it from buckling.

A tensioner is basically a piston inside a cylinder (fig. 91). The cylinder above the piston contains high-pressure air, and hydraulic fluid (a special lightweight oil) fills the cylinder below the piston. By increasing or decreasing the air pressure, crew members can adjust riser tension—more pressure, more tension, and vice versa. The hydraulic fluid dampens piston movement (makes it move more smoothly) and lubricates it.

BUOYANT RISER

When drilling in very deep water, the weight of the riser joints can become so great that the riser tensioner system alone cannot provide enough tension to support them. To overcome the problem, manufacturers attach buoyant devices—steel cylinders filled with air or plastic foam devices—to some

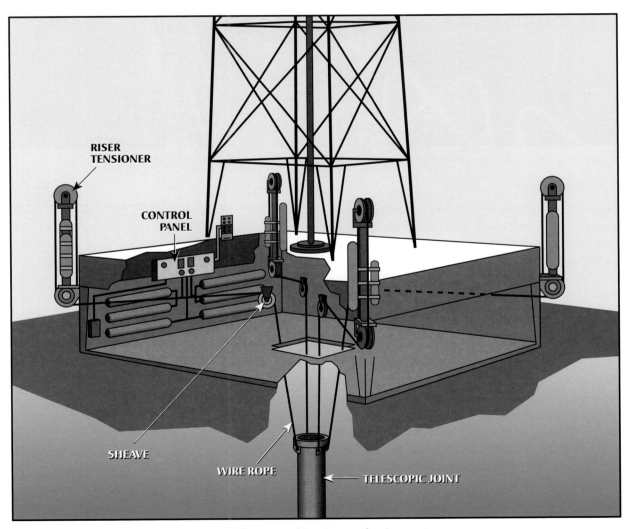

Figure 90.   A riser tensioning system provides upward tension on the riser system.

RISER
TENSIONER
ASSEMBLY

Figure 91. These three piston-and-cylinder assemblies are part of the riser tensioning system.

of the riser joints (fig. 92). Adding buoyancy to riser joints makes them weigh less when they are submerged and therefore allows the tensioner system to work effectively.

GUIDELINELESS METHODS

Another problem associated with drilling in very deep water concerns the guideline system. Guidelines are tensioned much as the riser system is, and similar problems can occur with guidelines in very deep water. To solve guideline problems, engineers have developed guidelineless methods of lowering subsea components to the ocean bottom.

Closed-circuit television, for example, enables the drilling crew to see what is happening while the driller lowers a component. The crew mounts the camera directly on the component, and a remote TV screen on the floater allows the driller to watch what is happening. Floaters also employ sonar systems, in which crew members mount sonar

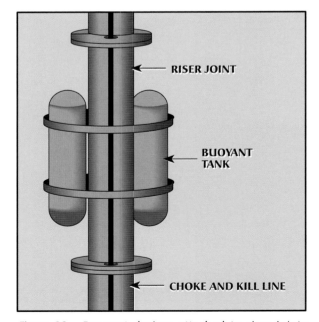

RISER JOINT

BUOYANT
TANK

CHOKE AND KILL LINE

Figure 92. Buoyant devices attached to riser joints augment the riser tensioner system.

A PRIMER OF OFFSHORE OPERATIONS

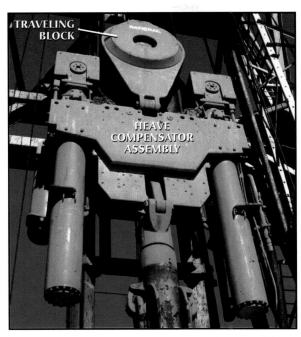

Figure 93. This heave compensator assembly is attached to the traveling block. Note the two pistons and cylinders that offset heave motion.

sending and receiving devices directly onto subsea components. What is more, personnel can use *remotely operated vehicles* (*ROVs*). They control these underwater robot vehicles from the rig to position subsea components.

HEAVE COMPENSATION

In any floating drilling operation, the rig has to compensate for heave on the drill stem; otherwise, weight on the bit can fluctuate with the vessel's heave. Therefore, crew members almost always install heave compensators, or drill string compensators, on floaters.

Heave compensators consist of a piston and a cylinder (or two pistons and cylinders) that are an integral part of, or are mounted directly on, the traveling block (fig. 93). As the floater moves up and down, hydraulic pressure within the piston-and-cylinder arrangement compensates for vertical movement of the block.

The piston strokes up and down within the cylinder as the traveling block moves, thereby keeping the hook in a fixed position relative to the seafloor. Since the drill stem hangs from the hook, the compensator offsets up-and-down movement (fig. 94).

To keep the block from moving side to side or in a circle, the rig builder usually mounts it in guide

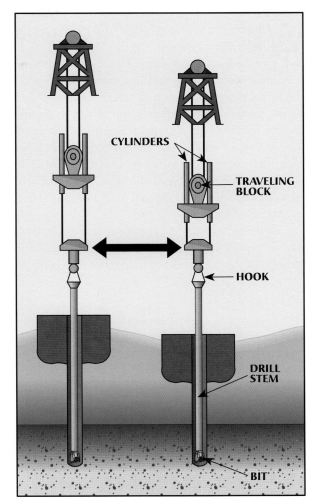

Figure 94. The heave compensator keeps weight on the bit at a constant value in spite of the floater's up-and-down movement.

tracks. These tracks, placed in the derrick on both sides of the block, allow the driller to raise and lower the block but minimize sideways or circular movements.

ADDITIONAL DRILLING AND CASING

Once crew members install the subsea BOP stack and the marine riser system, the rig can drill more hole. The crew lowers the bit and drill stem into the riser, through the BOP stack, and to the bottom of the conductor casing.

The driller starts the mud pump to begin circulating mud, rotates the bit and lowers it to bottom, and commences drilling. The crew will probably run and cement at least three more casing strings before the hole reaches the formation of interest.

Formation Evaluation

Whether the operator drills an exploration well from a bottom-supported or a floating unit, they must evaluate prospective formations. That is, the operator must determine whether the well has struck oil or gas in sufficient quantities to justify further reservoir development. Because the cost of developing, or exploiting, offshore reservoirs is so high, the quantity of oil or gas must be large. In fact, companies sometimes abandon reservoirs discovered offshore even though they contain a fair amount of oil and gas.

Evaluation involves measuring certain properties of a potential reservoir rock and obtaining samples of the rock and any fluids that may be in it.

WELL LOGGING

Well logging is an evaluation method in which a logging crew lowers a special tool, a *sonde*, into the well and then pulls it back up. As the sonde passes through the formations on its way up the wellbore, it senses and measures electrical, radioactive, and acoustic (sound) properties of the rocks.

Measuring and recording these properties provides a good deal of information about the rocks. The sonde transmits its measurements via conducting cable to the surface, where computers record them.

Experts analyze the resulting recording, the *log* (fig. 95). Analysis involves careful study of the curves on the log. The curves indicate whether hydrocarbons exist in the formations investigated and to what extent.

CORING

Coring is another method of formation evaluation. The operator obtains one type of core sample by lowering a *sidewall sampler* into the borehole. At the desired interval, the driller activates the sampler to fire several small cylinders into the wall of the hole. The cylinders penetrate a short distance into the rock and cut a small *core*. When crew members start to remove the tool from the hole, cables attached to the cylinders pull them from the wall of the hole, and the cores come up the hole with the tool. Geologists can then examine the small cores and analyze them for indications of the character of the formation.

Another method of coring involves replacing the regular drill bit with a special *core bit*. A core bit is shaped like a doughnut (fig. 96). The outside portion of the bit cuts formation, but since the center part of the bit is open, it leaves a core of rock in the middle.

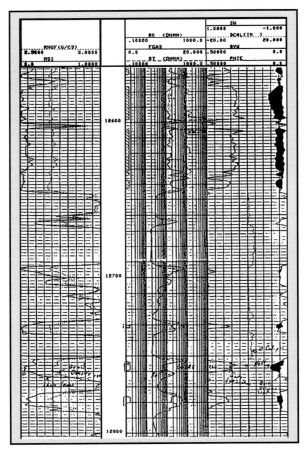

Figure 95. The curves on this well log can reveal information about the formation penetrated by the hole.

Figure 96. When made up on a core barrel, the core bit *(left)* cuts a long cylindrical sample of the formation, which is called a core *(right)*. Only a short portion of the core is shown.

A core barrel above the bit retains the core. When crew members pull the drill stem, core barrel, and bit back to the surface, specialists remove the core from the barrel and send it to a laboratory for analysis (fig. 97).

Figure 97.   Several cores are laid out for analysis at an onshore laboratory.

DRILL STEM TESTING

Drill stem testing is a method of formation evaluation. The test allows fluids in the formation being evaluated to flow into the drill stem and to the surface. Pressure-recording devices in the drill stem test tool (DST) measure and record pressure in the well while it is both flowing and not flowing (shut in). A DST can also obtain actual fluid samples from the formation. From detailed analysis of the pressure recordings and of the fluids, the operator can infer a great deal of information about the reservoir.

Since oil and gas may flow to the surface during a drill stem test, the operator must dispose of them properly. Commonly, special burners atomize the oil (break it into small droplets) and mix it and any gas with air, allowing for the relatively complete combustion of the hydrocarbons and minimizing air pollution (fig. 98).

Figure 98.   Special burners flare gas harmlessly into the atmosphere on this unit in the North Sea. *(Courtesy Marathon Oil Co.)*

Well Abandonment

Formation evaluation can reveal whether the exploratory hole has penetrated a reservoir or reservoirs that bear commercial quantities of hydrocarbons. Since the operator does not have immediate means to treat any oil and gas in an exploratory hole and transport them to market, the operator usually abandons the exploratory hole. The operator usually has to install a platform or a subsea production facility to exploit a productive reservoir.

Abandoning a well involves setting several cement plugs in it at various depths. These cement plugs permanently seal it up. The rig crew retrieves all the subsea equipment, and the operator cuts off and retrieves any casing or other equipment that extends above the mudline—the seafloor. Regulations usually require that no hazards to navigation be left behind on the seabed.

In some cases, especially in deep waters, operators may not abandon a productive exploratory well. Instead, they may temporarily cap it for later completion. Operators often drill deepwater development wells with MODUs. Special subsea production equipment is adaptable to subsea drilling equipment, thereby making the job of completion practical.

Summary

Drilling an exploration well is the only sure way to confirm the presence of hydrocarbons in a suspected reservoir. Operators usually drill an exploratory well with a mobile offshore drilling unit, either a bottom-supported unit, such as a submersible or jackup, or a floater, such as a drill ship or semisubmersible.

Drilling from floaters usually involves subsea BOPs, a marine riser system, and motion compensators. Drilling from bottom-supported units usually does not require a subsea system. Instead, crew members mount the BOPs on casing that projects above the water's surface. And since bottom-supported units firmly contact the seafloor, they do not require motion compensation.

Operators evaluate formations to determine whether sufficient amounts of hydrocarbons exist in the reservoir. Whether the first wildcat well yields hydrocarbons, it is usually, but not always, plugged and abandoned.

Should evaluation reveal promising results, the operator will probably drill more wells to confirm the findings. If all goes well, the operating company will go into full-scale development drilling to produce the hydrocarbons.

6
DEVELOPMENT DRILLING AND COMPLETION

If tests on an exploratory well prove favorable, the operator usually drills and evaluates several additional wells in or near the reservoir. These additional wells are *appraisal wells*. The operator drills appraisal wells to confirm further that the reservoir contains enough hydrocarbons to justify the enormous expense of developing it. If the appraisal wells reveal that the reservoir does indeed contain enough hydrocarbons, then development drilling may occur. *Development drilling* is the drilling of several wells into a reservoir

to extract hydrocarbons discovered by exploratory wells and confirmed by appraisal wells.

Operators use many kinds of rigs to drill *development wells*. A common type in U.S. waters is the platform rig, which may be rigid or compliant. A *rigid platform* does not move with the motion of the wind and sea; a *compliant platform* does move. Operators also use mobile offshore drilling units to drill development wells, the same rigs they often employ to drill exploratory and appraisal wells.

Drilling Platforms

In 1947, off the Gulf Coast of Louisiana, an operator drilled the first offshore well out of sight of land. It was an exploratory well, but the rig that drilled it was not, strictly speaking, mobile. Borrowing ideas and technology from experiences in Lake Maracaibo, engineers hit upon a sort of hybrid design for an offshore rig. It was a combination of a ship-shaped barge and a fixed platform. A *ship-shaped barge* is like a drill ship, except that it is not self-propelled. The operator has to tow it to the site. A *fixed platform* is a structure made of steel or concrete, although manufacturers originally made some out of wood. The operator firmly fixes it to the bottom of the body of water in which it rests.

The platform that drilled the well in 1947 was relatively small—only large enough to support the derrick, rotary table, and drawworks. The rig builders drove several steel piles and creosote-soaked timber piles into the seafloor. They then built a platform across the piles to provide a deck for equipment.

The barge was a U.S. Navy surplus landing craft converted to contain the rig's engines, mud pumps, mud pits, crew quarters, pipe racks, and so on. A

narrow walkway, a *widow-maker*, placed between the barge and the platform allowed the crew to move back and forth between the two. Operators soon began calling the barge a *drilling tender*.

The idea behind this combination barge-and-platform design was simple. After operators drilled the well from the platform, they could quickly tow the barge to a new location. Moreover, they could easily disassemble the small drilling platform and salvage most of it for use at the new site. Further, they could leave the basic platform standing and use it to house equipment needed to produce the well if they discovered significant quantities of oil and gas. The design proved to be a winner. Even today, you can still find drilling platform tenders (fig. 99).

As offshore drilling technology progressed, mobile offshore drilling units gradually replaced most platform tender rigs, especially for drilling wildcat (exploration) wells. Platform tenders established an important benchmark, however. They showed the feasibility of drilling development wells from a platform that an operator firmly attached to the seafloor.

Figure 99.   A platform tender is anchored next to a relatively small platform. *(Courtesy Sedco Forex Schlumberger)*

Today operators drill a large number of offshore development wells from totally fixed, self-contained platforms. The operator drills, completes, and produces all the development wells from the platform, thus eliminating the need for rig mobility. Modern platforms house all the drilling equipment, production equipment, crew quarters, offices, galley, and recreation rooms (fig. 100). The platform rigs of today vary in design and appearance. Manufacturers build some of steel and some of concrete. Some platforms are rigid; others are compliant. Rigid platforms include the steel-jacket platform, the concrete gravity platform, and the caisson-type platform.

STEEL-JACKET PLATFORMS

A *steel-jacket rigid platform* consists of a *jacket*—a tall, vertical section made of tubular steel members—which is the foundation of the platform. Piles driven into the seabed support it. The jacket extends upward so that the top rises above the waterline. Additional sections on top of the jacket provide space for crew quarters, a drilling rig, and all the

Figure 100. A self-contained platform houses all the drilling and production equipment, crew quarters, galley, offices, and recreation rooms.

A PRIMER OF OFFSHORE OPERATIONS

Figure 101. A platform jacket is so tall, it is usually built and transported on its side.

equipment needed to drill. The height of the jacket depends on the water's depth. Some of the tallest stand in water over 1,000 feet (300 metres) deep.

Most platform jackets are so large that builders construct them on shore, launch them into the water or place them on a barge, and tow them to the site. Usually, they build the jacket on its side, since it is so tall (fig. 101). The construction contractor seals several of the tubular members airtight so that when launched into the water, the jacket floats on its side, just as they built it.

At the production site, large barge-mounted cranes stabilize the structure, while an erection crew floods the tubular members (the legs). As the tubular members fill with water, cranes raise the jacket to a vertical position so that the legs come to rest on the seabed.

The construction crew then places steel piles through several of the jacket's legs and use a pile driver to drive the piles deeply into the ocean bottom. The piles pin the jacket firmly to the bottom and transfer loads from the platform to the seabed. Once the piles pin the jacket, cranes place the remaining platform elements on top of the jacket (fig. 102).

Figure 102. A large crane hoists a deck section onto the jacket. *(Courtesy Shell Oil Co.)*

Development Drilling and Completion

Figure 103. Five boats tow a concrete platform to a site in the North Sea.

CONCRETE GRAVITY PLATFORMS

In the North Sea, especially Norway's sector, rig builders make *concrete gravity rigid platforms* from steel-reinforced concrete. They construct them in deep, protected waters near shore. They then float and tow to the site in a vertical position (fig. 103). Some have three or four tall concrete caissons, or columns, that resemble smokestacks. At the site, the operator floods the caissons, and the platform submerges to come to rest on bottom.

One end of the caissons rests on the seafloor. The rig builders install drilling equipment, crew quarters, and production equipment on a platform on the other end, well above the waterline. Concrete platforms weigh a tremendous amount, so builders do not have to pin them to the bottom with pilings. The force of gravity alone keeps them in place.

Concrete platforms offer advantages over steel platforms. They can be constructed to store up to a million barrels of oil in special concrete cylinders arranged around their base on the seabed (fig. 104). Oil storage capacity is especially advantageous when no pipeline exists for transporting oil to shore. In such cases, tanker ships can tie up to the platform and load the oil for transportation. Concrete grav-

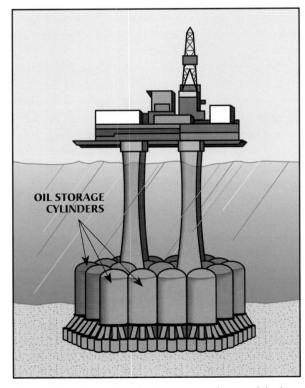

Figure 104. Concrete cylinders arranged around the base of a concrete gravity platform provide storage space for oil.

A PRIMER OF OFFSHORE OPERATIONS

ity platforms are also able to withstand extremely rough seas because of their tremendous weight. For this reason, operators use a lot of concrete gravity platforms in the North Sea.

CAISSON-TYPE PLATFORMS

Rigid platforms for certain arctic waters—the Cook Inlet of Alaska, for example—are constructed of steel but do not feature a jacket. Since moving pack ice can destroy a conventional jacket, engineers have developed platforms that stand on steel caissons.

One design has a caisson at each of the corners of the rectangular platform. The construction contractor firmly affixes the caissons to the seafloor, and lays the drilling and production decks on top of them (fig. 105). Since the operator drills development wells through the middle of each caisson, they use a drilling rig that they can move over the platform to each of the four caissons. The caissons also protect the top part of each well from moving ice.

COMPLIANT PLATFORMS

When drilling and production take place in deeper waters, rigid platforms are not feasible because they are very expensive to build and install.

Alternate designs are *compliant* rather than rigid. The platforms actually yield to water and wind movements, much as floating rigs do.

Guyed-tower platforms are similar to rigid-jacket platforms in that a construction crew pins the bottom of the platform (the jacket) to the seafloor. However, the guyed jacket is much slimmer and therefore does not contain as much steel as conventional jackets. It weighs less and is less expensive to build.

The rig builder attaches guy wires to the jacket fairly close to the water's surface and spreads them out evenly around the jacket. They then anchor the guy wires with *clump weights*. Clump weights allow the guy wires to move as the platform moves with wind and wave forces (fig. 106). The clump weights firmly anchor the guy wires and apply tension to them, yet still allow them to move as the platform moves.

Figure 105. A caisson-type platform rests in the ice-free waters of early summer in the Cook Inlet of Alaska. *(Courtesy Shell Oil Co.)*

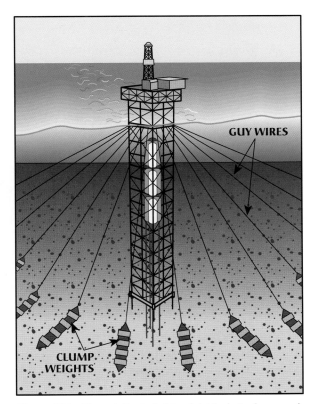

Figure 106. The relatively lightweight jacket of a guyed-tower platform is supported by several guywires and clump weights.

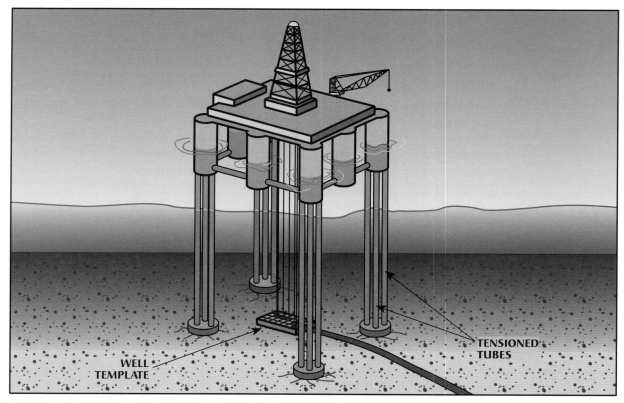

WELL TEMPLATE

TENSIONED TUBES

Figure 107. A buoyant tension-leg platform is held on station by several tensioned tubes attached to the seafloor.

Tension-leg platforms, like guyed-tower platforms, move with wind, waves, and currents. The rig builder attaches the platform, which resembles a semisubmersible drilling rig, to the seafloor with tensioned steel tubes. Since they firmly affix the tensioned tubes to the platform as well as to the seafloor, the buoyancy of the platform applies tension to the tubes (fig. 107).

Mobile Offshore Drilling Units

In some cases, an operator uses a MODU, such as a jackup, to drill development wells. For example, the operator can erect a relatively small platform jacket, only large enough to provide space for drilling wellheads, BOP stack, and, ultimately, production wellheads, on the site. In such cases, the drilling contractor places the MODU next to the jacket, and drills and completes the wells on it (fig. 108).

After the wells are drilled, the contractor moves the MODU away and, if necessary, the operator erects an additional platform nearby to house production equipment and quarters for the production crew (fig. 109). A system of two relatively small platforms and a MODU can reduce costs below those

of a single large drilling and production platform.

Operators can also use mobile rigs to drill development wells into a reservoir, perhaps into parts of it that are not accessible from a platform located over another section. They drill these *satellite wells* one at a time; that is, the MODU drills one well and then moves to drill other wells at different locations over the reservoir. This method of drilling a single development well and then moving the rig is practical only with MODUs. Operators employ a different technique when they drill development wells from platforms.

Sometimes it is not economically efficient to drill several single development wells at various points into a reservoir. In such cases, operators can

Figure 108. A jackup drills development wells on a small jacket *(right)*.

Figure 109. Two relatively small platforms replace a single large platform. *(Courtesy Shell Oil Co.)*

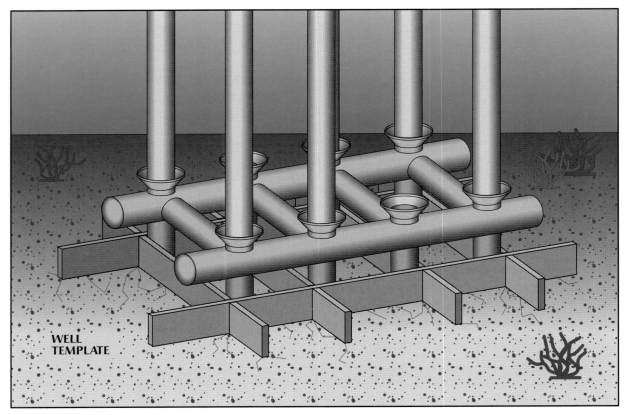

Figure 110. A subsea template allows several wells to be drilled, completed, and produced from a relatively small area on the seafloor.

use MODUs with *subsea templates*. A subsea template is an arrangement of special equipment the operator places on the seafloor. The template allows the operator to drill, complete, and produce several wells from the relatively small point on the seafloor that the template covers (fig. 110). The operator places the MODU, usually a floater, on the water's surface over the template, and drills and completes several wells on the template.

Directional and Horizontal Drilling

Operators drilling development wells from platforms, or from MODUs using a subsea template, usually employ a special drilling method. Remember that even though a platform can be a huge structure, it occupies only a relatively small area with respect to the area of the reservoir.

A single well drilled into most offshore reservoirs cannot effectively drain the reservoir of hydrocarbons. Several wells, perhaps forty or more, drilled into many points in the reservoir may be required to produce it. Yet the platform is not movable. How can the operator drill so many wells from a single, unmovable platform? The answer lies in a special drilling technique.

When operators drill most offshore wildcat wells, they make every effort to drill the hole as straight as possible. After all, the shortest distance between two points is a straight line. But straight holes are not the answer when it comes to drilling development wells from a platform.

Instead, operators use a special drilling technique, *controlled directional drilling*. They drill the top part of the well as straight as possible, then deflect it off vertical so that the wellbore curves in the desired direction. Using special drilling tools and devices that measure the direction and angle of the hole, the offshore directional driller is able to guide each wellbore into a different part of the reservoir (fig. 111).

Usually, operators spud each well so that the top part of each runs parallel to the others. At a predetermined depth, they deviate the wells to different parts of the reservoir. The effect is not unlike a huge spider crouched over the reservoir, with each leg being a well to a different section of the reservoir.

Another technique is horizontal drilling. In this case, the operator deflects the well off vertical to the point that it runs parallel to the surface. That is, operators deflect horizontal wells 90 degrees (or more) from vertical. They can produce some reservoirs much more effectively with a single horizontal well than they can with several vertical wells.

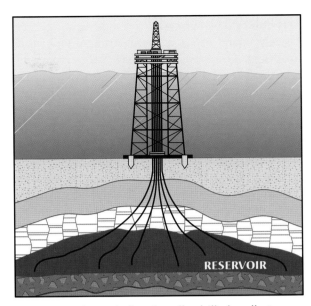

Figure 111. Several directionally drilled wells tap an offshore reservoir.

Well Completion

After operators drill a development well into the reservoir and run and cement the final string of casing, they must *complete* the well. The operator must provide a way for the reservoir fluids to enter the well and reach the surface. Engineers have developed many types of completion for both on land and offshore, but we'll cover only the most common one.

PERFORATING

Picture a well drilled through a reservoir. Casing and a surrounding sheath of cement line the well. The hydrocarbons in the reservoir cannot flow into the well because the cement and casing block their entry. To open the well to the reservoir, the operator can *perforate* the casing and cement. That is, a perforating company can make several small holes in the casing and cement. Hydrocarbons flow from the reservoir into the well through the perforations.

To perforate a well, a perforating operator lowers a special perforating gun to a point opposite the producing formation. The operator fires the gun, and several small, but intense, explosions penetrate the casing and cement and travel a short distance into the reservoir. These small explosions are *shaped charges*—explosives that release a high-energy charge of gases and particles (fig. 112). These gases and particles travel in one direction and penetrate casing, cement, and formation. Reservoir fluids then enter the well through the perforations.

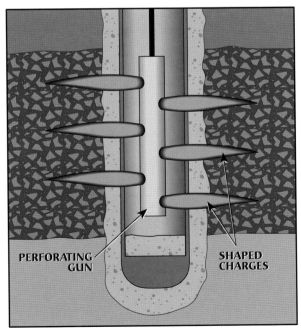

Figure 112. A perforating gun is lowered into the wellbore and fired. Shaped charges perforate the casing, cement, and formation.

TUBING AND PACKERS

Usually, operators run a special string of pipe inside the casing. This special pipe is *tubing*. Reservoir fluids flow up the well through the tubing that is inside the cemented casing. Operators produce most wells through tubing instead of directly through the casing. They produce through tubing for several reasons. For one thing, the crew does not cement the tubing in the well. As a result, when a joint of tubing fails, as it almost inevitably will over the life of the well, the operator can easily replace the failed joint. Since casing is cemented, it is very hard to replace.

For another thing, tubing allows the operator to control the well's production by placing special tools and devices on or in the tubing string. These devices allow operators to produce the well efficiently. Casing does not provide a place to install tools or devices required for production. In addition, the operator usually installs safety valves in the tubing string. These valves automatically shut off the flow of fluids from the well if damage occurs at the surface.

Finally, tubing protects casing from the corrosive and erosive effects of produced fluids. Over the life of a well, reservoir fluids tend to corrode metals with which they are in contact. Producing fluids through tubing, which the operator can easily replace, preserves the casing, which is not so easy to repair or replace.

Crew members often install a *packer* near the bottom of the tubing above the producing zone. A packer is a length of pipe through which well fluids flow. The packer has gripping elements that latch onto the inside wall of the casing to help anchor the packer. Rubber sealing elements on the packer, when expanded, provide a pressure-tight seal between the tubing and casing. The packer forms a fluid-tight seal between the tubing and the casing. Thus, when reservoir fluids flow into the well, the packer forces them to flow into the tubing (fig. 113).

SURFACE AND SUBSEA COMPLETIONS

Completing a well involves the control of fluids as they flow from the top of the well. The operator can place the devices that control fluid flow either above the surface of the water or onto the seafloor. In a *surface completion*, the operator places the equipment that controls the flow of hydrocarbons from the well above the waterline. Operators usually complete the wells they drill from a platform on the platform itself (fig. 114). In a *subsea completion*, the operator

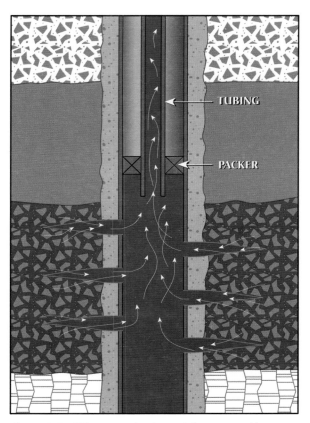

Figure 113. When a packer is set, it forms a seal between the outside of the tubing and the inside of the casing, causing reservoir fluids to flow into the tubing.

places the equipment that controls the flow of hydrocarbons from the well on or below the seafloor.

In a subsea completion, the operator locates the collection of special valves and fittings that controls the flow of hydrocarbons from the well—the *Christmas tree*—at the top of the well on the seafloor. A subsea Christmas tree can be wet or dry. If wet, the operator installs it on the wellhead where it remains exposed to the surrounding water (fig. 115). If dry, the operator installs the Christmas tree in some type of housing that isolates it from the water, often at atmospheric pressure (fig. 116).

Divers may maintain and repair wet trees; sometimes, the operator utilizes remotely operated vehicles (ROVs) to make repairs. Operators can maintain some dry trees with small submarines called *submersibles* or *diving bells* to gain access to the housing around the Christmas tree. Workers enter the housing without any special clothing or underwater breathing equipment.

Figure 114. In a surface completion, equipment that controls the flow of hydrocarbons from the well is placed on a deck of the platform. *(Courtesy Shell Oil Co.)*

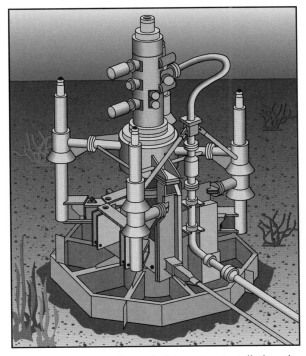

Figure 115. A wet subsea Christmas tree installed on the seafloor

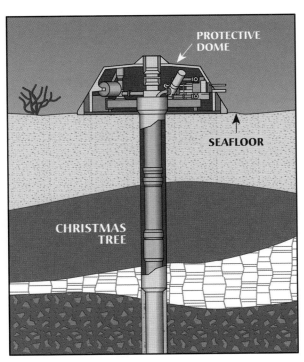

Figure 116. This dry subsea Christmas tree is isolated from the water by a protective dome.

Summary

In offshore development drilling, operators often drill several wells from a fixed, self-contained platform. They usually drill these development wells directionally so that each wellbore curves into various parts of the reservoir. Sometimes, operators use horizontal wells to tap into a reservoir.

Some rigid platforms consist of steel jackets that piles driven into the seafloor support; other rigid platforms are made of concrete. Concrete platforms usually do not require piles because they are so heavy. In the Arctic, operators sometimes use special caisson-type platforms to prevent the destruction of the platform by moving pack ice.

When operators drill and complete wells in deeper waters, they often employ compliant platforms that move to a limited extent with wind, waves, and currents. One design is a guyed tower, in which several guy wires are attached to a relatively lightweight jacket and to the seafloor. Another is the tension-leg platform, in which steel tubes in tension are firmly anchored to the seafloor and platform, keeping the floating platform on the site.

Sometimes, operators may use a mobile offshore drilling unit to drill development wells. In such cases, they place the MODU next to a small platform jacket, and drill the wells through it. If necessary, a small additional platform is erected next to the original platform to house equipment that cannot fit onto the original.

In some instances, operators drill satellite wells to exploit a reservoir. They drill satellite wells one at a time with a MODU. Operators often use satellite wells to supplement development wells drilled from nearby platforms. Subsea templates and directional drilling allow floating rigs to drill several development wells without having to move from well site to well site. The template provides a way for the floater to drill, complete, and produce several wells in the small area occupied by the template.

Well completion usually involves perforating the well to make holes in the casing and cement. The perforations permit reservoir fluids to enter the well. The operator usually places a packer with the tubing, and reservoir fluids flow up the well through the tubing.

Completions may be either surface or subsea. If the operator places the Christmas tree above the waterline, the completion is a surface completion. If the operator places the Christmas tree on or below the seafloor, the completion is a subsea completion.

7
PRODUCTION AND WORKOVER

Offshore production involves a wide range of techniques and equipment. The operator uses the techniques and equipment to get oil out of the reservoir and into a transportation system. As produced fluids flow up the well and to the surface, the operator has to separate oil, gas, and water, and must handle each according to its individual requirements. As a result, a typical production platform may have a large amount of equipment and several people to successfully extract, treat, and move oil and gas to shore for refining and processing.

Reservoir Drive Mechanisms

Natural pressure in the reservoir forces oil and gas to flow from a reservoir into a well and to the surface. The main sources of this pressure are gas, water, or both in combination. Four kinds of natural drive mechanisms occur with oil and gas: dissolved-gas drive, gas-cap drive, water drive, and combination drive.

Dissolved-gas drive is present when all or almost all of the hydrocarbons in the reservoir are liquid in its undrilled state. When the operator drills a well into such a reservoir, the well creates an area of reduced pressure. It is somewhat like jamming a straw into a bottle of liquid and sucking on the straw. Sucking on the straw reduces pressure and liquid flows up the straw. Similarly in a reservoir, reducing the well's pressure causes some of the lighter liquid hydrocarbons to become gaseous. The gas comes out of solution and, as it does, it drives oil into the well and to the surface (fig. 117).

Gas-cap drive results when a reservoir contains so much gas that all of it is not dissolved in the oil. Because gas is lighter than oil, it usually rises to the top of the reservoir and forms a cap over the oil. When the operator completes a well into such a reservoir, the gas cap expands. The expanding gas drives oil into the well and to the surface (fig. 118).

Water drive occurs when large amounts of water—usually salt water—lie beneath the oil in a reservoir. When the operator drills

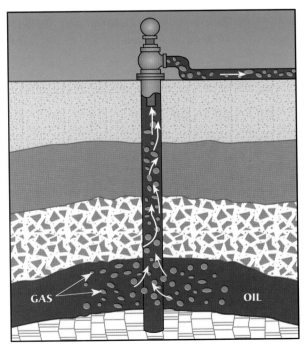

Figure 117. In a dissolved-gas drive reservoir, gas comes out of solution from the oil and drives oil to the surface.

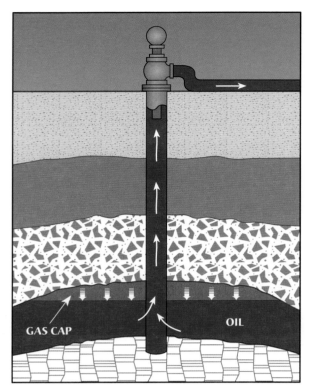

Figure 118. In a gas–cap drive, the gas cap expands to lift oil to the surface.

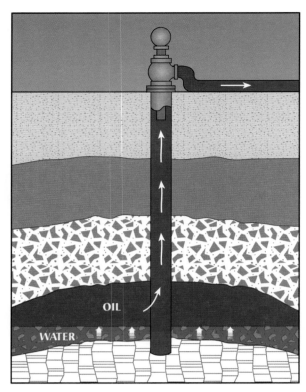

Figure 119. In a water drive, water moves oil to the well and up to the surface.

a well, the water in the reservoir moves toward the well and drives oil ahead of it (fig. 119).

Combination drive exists when both gas and water provide energy to drive oil to the surface. Gas—in the form of a cap or dissolved in the oil or a combination of the two—and water lying below the oil work to lift oil to the surface (fig. 120). Combination drive is the most common drive mechanism.

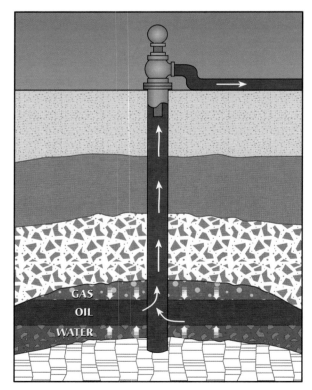

Figure 120. In a combination drive, expanding gas and water lift oil to the surface.

Handling Oil, Gas, and Water

Oil and gas seldom reside in a reservoir by themselves. Water and other impurities like sediment usually coexist with them. As operators (producers) produce the reservoir, these impurities come to the surface with the oil and gas. Operators must separate and remove the foreign material from the hydrocarbons. They must also separate the oil and gas from each other before they can transport them to shore for refining and processing.

Operators must separate and remove impurities—especially *sediment and water* (*S&W*)—because the impurities take up space in a tanker or pipeline that valuable hydrocarbons could otherwise occupy. Furthermore, water can corrode practically any steel surface, including that of a tanker compartment or a pipeline.

Operators have to separate oil from gas primarily because each has different handling requirements. For instance, the operator usually has to put gas under additional pressure to put it into a pipeline or special tanker for transport to shore facilities. Because oil is a liquid it cannot be compressed, and therefore must be pumped into a pipeline or tanker.

To separate and handle oil, gas, and water, operators use several techniques and a variety of equipment. They usually locate the equipment on the same platform from which they drilled the wells. But once they start producing the wells, the platform becomes a *production platform* and operators use the derrick for such purposes as raising and lowering equipment to work over the well. Because it's not being used strictly for drilling at this time, the derrick can be used to run special tools into wells that are producing but that need cleanout or repair (fig. 121).

In a subsea completion, the operator routes oil, gas, and water into flow lines that run from the wells to a *production riser*. Production flows up the riser to a floating buoy. An oil tanker can then tie up to the buoy, take on the produced fluids, and transport them

Figure 121. This production platform retains the derrick used in drilling the wells. (*Courtesy Petroleos Mexicanos*)

to shore for separation and treatment (fig. 122). In some cases, the operator moors a processing and storage ship to the buoy. Equipment on the ship separates, treats, and stores the reservoir fluids. When oil fills up the ship's tanks, the ship transports it to a receiving terminal on shore. Operators sometimes use this system in deep waters where erecting a platform is not economically feasible.

In shallower waters, operators can send production from subsea wells to a platform. Once on the platform, equipment separates the fluids and off-loads the oil onto a tanker, or injects it into a pipeline.

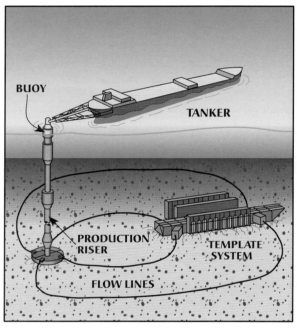

Figure 122. In this subsea production system, oil flows from wells on the template, up the production riser, and to a special mooring buoy, where a tanker loads the oil.

HANDLING PRODUCED WATER

Water produced with oil and gas occurs in three ways: (1) as free water; (2) combined with oil in an emulsion; or (3) mixed with gas as vapor.

Free-Water Knockouts

Free water is water not tightly bound to the oil and gas. It readily settles out if the operator gives the produced fluids a chance to remain stationary for a short period in a tank or vessel. Thus, one type of equipment producers sometimes use on an offshore production facility is a *free-water knockout* (*FWKO*) (fig. 123). A free-water knockout is usually a

Figure 123. A free-water knockout removes free water from produced fluids.

cylindrical vessel. The operator pipes well fluids that contain relatively large amounts of free water after they leave the Christmas tree into the FWKO. An FWKO can be either vertical (standing upright) or horizontal (lying on its side). Vertical designs take up less space but can't handle as much volume as horizontal ones. Horizontal vessels expose the fluids inside them to more surface area. The design selected depends primarily on the volume, or quantity, of the fluids being produced and the amount of space available on the platform.

When well fluids go into a free-water knockout, the free water, since it is heavier than oil or gas, falls to the bottom of the vessel. The operator then removes it and disposes of it properly. If significant amounts of gas are present in the fluid stream, the manufacturer can equip the knockout to handle the gas. In such cases, the gas rises to the top of the knockout. The producer removes the gas and sends it to other equipment on the platform for further handling. Oil generally winds up in the middle. The operator pipes the oil to equipment on the platform for further handling.

Treaters

Sometimes, water tightly binds to the oil in the produced fluids. Usually, this water spreads out (disperses) in the oil as small droplets that cannot readily settle out. This combination of oil and water is an *emulsion* (fig. 124).

The operator specially treats emulsions to remove the emulsified water from the oil. To treat an emulsion, producers use special vessels—*treaters*.

Figure 124. A photomicrograph of a water-in-oil emulsion reveals the small droplets of water that are dispersed through the oil. *(Courtesy Shell Oil Co.)*

They pipe the emulsion into the treater and use chemicals, electricity, heat, or all three to break the emulsion. Breaking the emulsion means making the water settle out of the oil.

Usually, the operator adds chemicals to the emulsion and pipes it into the treater. Treaters, like free-water knockouts, can be either vertical or horizontal (fig. 125). Chemical companies formulate the chemicals to make the small water drops merge into large drops. Large water drops fall from the oil to the bottom of the treater where the operator removes them.

Electricity can also cause small water drops to merge and fall to the bottom of a treater. If an emulsion, usually with emulsion breaking chemicals in it, passes over a high-voltage electric grid in the treater, the water droplets merge and fall. *Electrostatic treaters* use electricity to treat, or break, an emulsion.

Heat reduces the viscosity, or thickness, of oil. Since water can fall out of a heated, thin oil more easily than it can out of a cool, thick oil, operators sometimes use *heater-treaters* in the emulsion-breaking process. In such cases, a *fire tube*, which is an enclosed gas flame, heats the emulsion while it is in the vessel. Operators often use chemicals and electricity in addition to heat.

Figure 125. A horizontal treater is used to remove most of the emulsified water from oil.

Glycol Dehydrators

Water may occur with natural gas as vapor. *Water vapor* is water in its gaseous state. Producers have to remove most of the water vapor before they can put the natural gas into a pipeline for transportation to shore. To put it another way, producers must *dehydrate* the gas; otherwise, the water vapor in it can liquefy, take up space in the pipeline, cause corrosion, and create other problems. For example, water in gas can form icelike solid substances—*hydrates*—that restrict or block the flow of gas in a line.

To dehydrate natural gas, operators often use a *glycol dehydrator* (fig. 126). *Glycol* is a liquid that has an affinity for water vapor. When gas with water vapor in it—*wet gas*—bubbles through the glycol in the dehydrator, the glycol dries the gas by absorbing most of the water. The dehydration equipment then pipes glycol that has absorbed water in it—*wet glycol*—to a special heater, a *reboiler*. The reboiler boils off the water, leaving *dry glycol* behind. The dehydrator then reuses the dry glycol to dry wet gas.

Water Disposal

Once producers remove water from oil and gas, they must properly dispose of the water. Offshore, proper disposal can involve putting it into the sea. Before operators do this, however, they treat the water in equipment that removes virtually all the oil.

Sometimes, the producer disposes of water by pumping it back into a suitable underground formation through wells set up especially for this purpose. In some cases, injecting water back into the subsurface can help drive additional amounts of oil into producing wells, much as a natural water drive does.

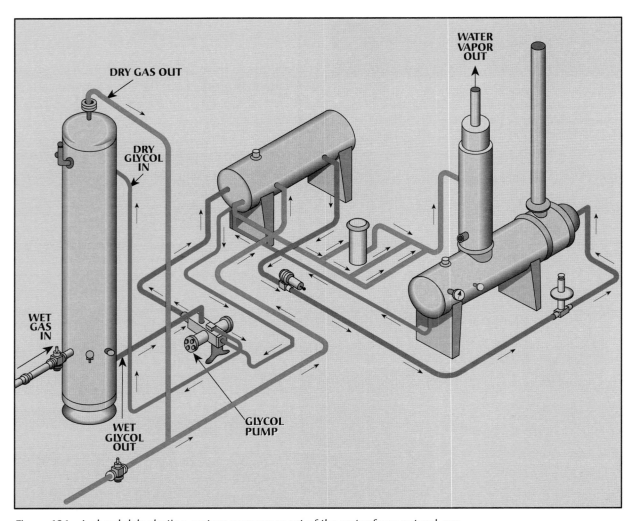

Figure 126. A glycol dehydration system removes most of the water from natural gas.

Figure 127. Two horizontal separators installed on this platform separate oil and gas.

SEPARATION OF OIL AND GAS

To separate oil and gas from one another, the producer usually employs a *separator* (fig. 127). Most separators are vertical or horizontal, although a spherical (ball-shaped) design is available. All provide an enclosed space into which the operator pipes produced fluids. Given a chance to remain stationary for a short period, the fluids tend to separate.

Some separators are two-phase, which means that only oil and gas or emulsion and gas separate inside them. Other separators are three-phase, which means that oil (or emulsion), gas, and water separate inside them. If only relatively small amounts of free water are in the produced fluids, or if some free water remains after the fluids pass through a free-water knockout, then the producer may use a three-phase separator.

As separation occurs inside the separator, the gas goes to the top of the vessel. The operator pipes it to additional gas-handling facilities on the platform—to a glycol dehydration system, for example. If water separates out, operators remove it from the bottom and dispose of it. They remove oil from the middle (or bottom, if they are using a two-phase separator) and transport it to shore if the oil does not require further treatment.

Usually, however, water is emulsified in the oil, so the operator pipes this emulsion to treaters or heater-treaters elsewhere on the platform. After the treaters break out most of the emulsified water, the operator transports the treated oil to shore.

Artificial Lift

As natural reservoir pressure produces hydrocarbons, this natural pressure declines. Eventually, the pressure dwindles to the point that it can no longer lift oil to the surface. The remaining pressure is not high enough to do the job. Natural reservoir pressure usually depletes long before the operator is able to extract all the hydrocarbons from the reservoir. So operators turn to *artificial lift* to produce at least some of the remaining hydrocarbons.

In artificial lift, the operator applies some form of external energy to the well. The applied energy must reduce pressure at the bottom of the well. It must reduce the bottomhole pressure enough so that the small amount of remaining reservoir pressure can force more reservoir fluids into the well and lift them to the surface. Offshore, the most common form of artificial lift is *gas lift*.

In gas lift, compressors inject gas into a well. This gas lowers pressure at the bottom of the well. To understand the method, imagine a well full of liquids (fluids) from the reservoir. The fluids exert pressure at the bottom of the well, just as a swimming pool full of water exerts pressure on the bottom of the pool. The pressure the well fluids exert at the bottom of the well is exactly the same as the pressure in the reservoir. Since fluids can flow only from an area of high pressure to an area of low pressure, no fluids can flow out of the reservoir.

Now picture injecting gas to the bottom of the well. Gas is a very lightweight, or low-density, fluid. A low-density fluid like gas does not exert as much pressure as a liquid. Injecting gas to the bottom of the well forces most of the liquids out of the well so that a mixture of liquid and gas remains. This liquid-gas mixture is lighter than the original liquid column and thus exerts less pressure on the reservoir at the bottom of the well. Since reservoir pressure is now higher than pressure inside the well, fluids from the reservoir flow into the well and to the surface.

In practice, gas injection is a little more complicated. Operators use compressors to inject gas into the well (fig. 128). The compressors must put

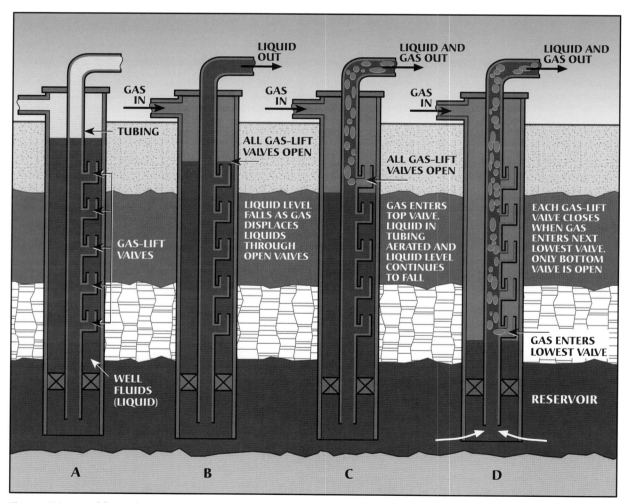

Figure 128. Gas lift involves the injection of gas into a well to lower pressure at the bottom of the well. (A) Since the well is not producing, no liquids are flowing; (B) injected gas forces liquid out through tubing; (C) gas enters top valve and lightens liquid in tubing, causing liquid level to fall further; (D) well is producing fluids from the reservoir.

enough pressure on the gas so that it can move down the well and force liquids out. Most wells are so deep, however, that extremely high gas pressure would be required to force all of the liquid out of a well in one step. So, operators usually inject gas in stages from the top of the well downward.

The operator installs several *gas-lift valves* at various well depths. All the valves are open at first. The compressor injects gas into the open top valve, which unloads (empties) the liquid from the top of the well. *Unloading* the liquid reduces pressure in the well to a point low enough that the injected gas can move down to the next valve. The first valve then closes and gas enters the second valve and further unloads the well. The gas moves down to the open third valve while the second valve closes. This progressive unloading of liquid continues until the gas reaches the bottommost gas-lift valve. At this point, the gas reduces well pressure, and reservoir fluids flow to the surface.

Additional Recovery Techniques

Artificial lift cannot produce all of the hydrocarbons that reside in a reservoir. Operators usually leave behind a fair amount of oil even after they apply artificial lift. They don't do it on purpose. It's a matter of the laws of physics and the behavior of fluids in a reservoir. In some reservoirs, operators leave up to 75 percent of the hydrocarbons originally in place. On the average, they can produce only about 33 percent of the total amount of oil in the reservoir. In an attempt to recover at least some of the remaining oil, companies have been conducting a great deal of research in recent years, and a few processes have shown promise. These processes go by various names: enhanced recovery, secondary recovery, tertiary recovery, and improved recovery. Here, we group them as "additional recovery."

One of the more traditional additional recovery methods is water or gas pressure maintenance. And in recent years operators are employing miscible flooding.

PRESSURE MAINTENANCE

Sometimes an operator can inject water into wells that penetrate the producing reservoir to recover additional oil. Operators may drill new wells to serve as water injection wells, or they may inject water down existing wells that are no longer producing. In either case, they introduce water into reservoirs that have little or no remaining water drive. The injected water behaves much like natural water drive and moves oil to producing wells (fig. 129).

Operators can reinject natural gas produced from a reservoir but for which no pipeline or other method of transport to shore is available. Reinjected gas may behave like gas-cap drive: the gas expands to force additional quantities of oil to the surface. If no method of transportation is available for gas, operators can also reinject it back into the reservoir and store it there until such time as transportation facilities are constructed.

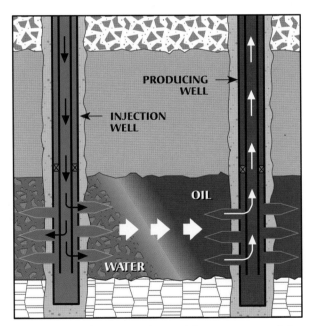

Figure 129. When water is used for pressure maintenance, it is injected into the reservoir to drive oil to producing wells. Some of the oil will be left behind, however.

MISCIBLE FLOODING

Even after water or gas injection, large quantities of oil may still remain in a reservoir. In recent years, operators have found that *miscible flooding* can recover more oil.

Miscible simply means capable of mixing. In miscible flooding, the operator injects a substance into the well that mixes with the oil remaining in the reservoir. This mixing sometimes enables additional amounts of oil to move to producing wells. For example, the operator can inject a *surfactant*—a chemical similar to detergents used to wash clothes—into a reservoir. The surfactant washes through the reservoir, mixes with some of the oil still locked in the rock pores, and releases it so that it can move into producing wells (fig. 130).

Another type of miscible flooding involves the use of *carbon dioxide* gas (CO_2). CO_2 mixes with some of the remaining oil, unlocks it from the pores, and the oil and CO_2 mixture move to producing wells. The operator often injects water behind the CO_2 to help prevent CO_2 from rising to the top parts of the reservoir (fig. 131).

Water and gas injection and miscible flooding are some of many additional recovery techniques that producers have tried or are using at the present time. Operators have tested and used these techniques mostly on land locations, since most older fields are on land and are the best candidates for additional recovery methods. As time passes and operators develop more offshore oilfields, these

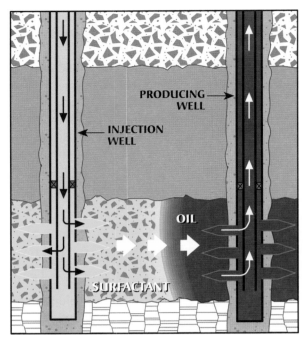

Figure 130. Injecting a surfactant can recover additional amounts of oil from a reservoir.

fields will also reach the limits of production by conventional means and they too will become targets for additional recovery techniques.

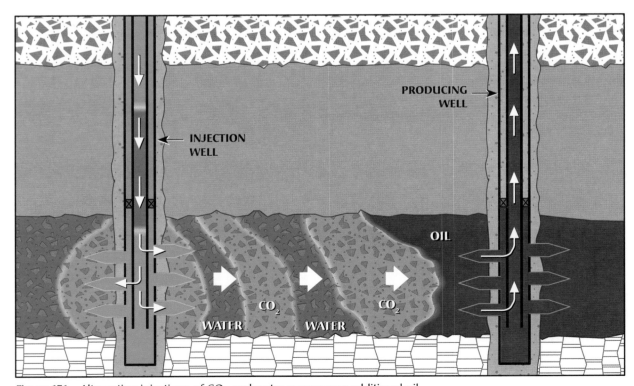

Figure 131. Alternating injections of CO_2 and water can recover additional oil.

A PRIMER OF OFFSHORE OPERATIONS

Well Servicing and Workover

As wells produce and start aging, a lot of negative things can happen to them. Downhole equipment wears out, sand from the reservoir gets into the well and restricts fluid flow, gas-lift valves malfunction, corrosion eats away metal parts—the list is long. Whatever the problem, anything that causes production to decrease or, in the worst case, to cease altogether is certainly going to get an operator's attention.

In some cases, the operator will hire a *well service and workover contractor*, a company that specializes in offshore well repair. In other cases, the operator will turn to the company's own employees who have the expertise and equipment to do it themselves. In either case, the goal of well servicing and workover is to fix whatever is wrong and restore production to a normal rate or to a rate as near normal as possible.

TUBING REPAIR

Operators and contractors use several techniques, tools, and methods to repair or maintain wells. On large, self-contained platforms, the operator often retains the derrick used for drilling the original wells and uses it for servicing those same wells. For instance, if part of the tubing string fails—maybe corrosion eats away part of it—a repair crew has to pull the entire string out of the well. They then replace the damaged portion and run the string back in. Having the original derrick in place makes pulling and running tubing relatively easy.

On the other hand, if the platform is small and has no derrick or if a satellite well needs to have its tubing pulled, then the operator may move in a mobile workover rig (fig. 132). Both jackup and semisubmersible workover rigs are available for servicing and workover.

WIRELINE METHODS

Wireline can run, manipulate, and pull much of the equipment operators place in the well when they complete it. *Wireline* is strong, fine wire that spools onto a reel (fig. 133). Repair crews can attach various tools and devices to wireline and lower or pull it from the well in one continuous movement. Using wireline is much faster than running and pulling a tubing string. Crew members make up tubing in joints much like they make up drill pipe. Since wireline is one continuous strand, crew members do not need a derrick to run tools and equipment on it.

One example of a wireline job is to remove and replace a gas-lift valve. If the operator installed the valve in the well's tubing in a special *side-pocket mandrel*,

Figure 132. A mobile jackup rig services a well. *(Courtesy Sundowner Offshore Services, a Nabors Industries Company)*

Figure 133. A wireline unit is used to make well repairs. *(Courtesy Shell Oil Co.)*

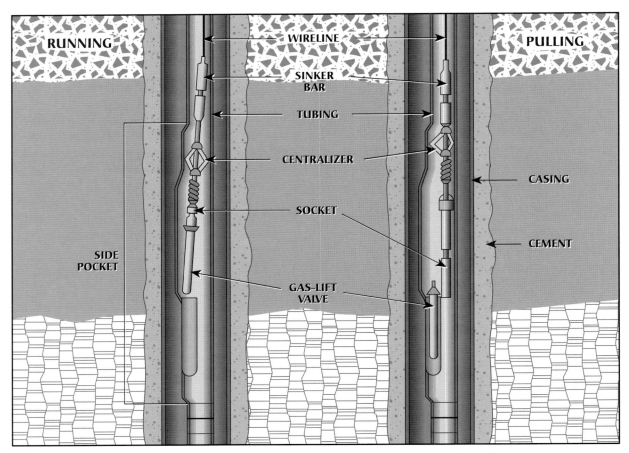

Figure 134. Gas-lift valves installed in side-pocket mandrels can be run and pulled with wireline.

the repair crew can retrieve and replace the valve on wireline (fig. 134). They attach a weight—a *sinker bar*—and a socket on the end of the wireline. They lower the assembly to the faulty valve and latch onto it. Crew members then pull the valve out of the side pocket, bring it to the surface, repair or replace it, and lower it back into the pocket. Crew members then release the socket from the valve, and reel the wireline to the surface. The process is a lot faster than pulling tubing to repair or replace a valve.

PUMPDOWN METHODS

For servicing subsea completions, repair crews can pump special tools down the flow line and into the tubing string to do a specific job. After they complete the job, they circulate the tool back to the surface. Such *pumpdown tools* are often called *through-the-flow-line tools*, or *TFL* tools for short. TFLs have simplified remedial work on subsea completions. Also, operators use them on surface completions as well.

Summary

Natural energy is usually present with hydrocarbons in the reservoir. This natural energy often comes from the gas and water that exist with oil. Gas, water, or both drive oil into wells that are drilled into the reservoir.

Operators must remove water and other impurities like sediment from oil and gas. Otherwise, they cannot put the oil or gas into a pipeline or tanker for transportation to refineries and processing plants on land. Refinery and plant equipment cannot handle excessive amounts of impurities. The operator uses several special tanks on an offshore production facility to separate and remove water from oil and gas. In some cases, especially where producers use subsea

A PRIMER OF OFFSHORE OPERATIONS

completions, they moor a processing and storage ship to a buoy. Well production flows up a production riser to the buoy. Separation and treating occur on the ship.

If free water exists in the fluids produced from the well, the producer may use a free-water knockout. A free-water knockout is a tank into which the producer pipes well fluids. It serves as an enclosed space where the free water gets time to settle out of the produced fluids.

Emulsified water requires additional treatment to remove it from oil. Producers apply chemicals, electricity, heat, or all three to the emulsion inside vessels known as treaters, or heater-treaters, if they use heat. Chemicals and electricity cause the small droplets of emulsified water to merge, become larger, and settle to the bottom of the treater. Often, the operator heats the emulsion because water drops can fall through a heated, thin oil faster than through a cool, thick oil.

To remove water vapor from gas, operators usually employ some type of dehydration system. The most common dehydrator uses glycol. Since glycol absorbs water from gas, glycol dehydrators bubble gas through glycol to dry it. The dehydrator heats the wet glycol to boil off the water. With the water gone, the dehydrator reuses the glycol to dehydrate more gas.

Once equipment removes water from hydrocarbons, operators must properly dispose of it. Sometimes, they put it into the surrounding seawater after they treat it to remove virtually all oil that may remain in it. In other cases, the operator may inject it back into the reservoir or other suitable subsurface formation for disposal.

Separators are tanks that separate gas from oil or gas from emulsion. Other separators separate gas, oil (or emulsion), and water. Operators have to separate gas and oil because each requires different handling on the platform. Also, they have to transport each in its own pipeline or tanker.

Gas lift is the most common form of offshore artificial lift. The producer injects gas into the well through special downhole valves, one at a time. Ultimately, pressure inside the well goes down low enough to allow the remaining reservoir pressure to lift fluids to the surface.

Operators may apply additional recovery techniques to recover more oil from a reservoir after conventional production techniques can no longer do the job. One technique is to inject water or gas into the reservoir. The water and gas maintain reservoir pressure. Another is miscible flooding, whereby the operator injects a chemical, such as CO_2, that mixes with the oil. Such techniques move oil into wells that the operator cannot produce by ordinary means.

Well servicing and workover is a vital part of offshore work. Operators must maintain and repair wells to keep them producing at the desired rate.

8

OIL AND GAS TRANSPORTATION

After operators produce, separate, treat, and, in some cases, store oil and gas on an offshore production facility, they must ultimately send them to shore. Once on land, refineries and processing plants make products that practically everyone in the world uses. In general, companies transport oil and gas in two ways: in pipelines and in tanker ships.

Transportation by Pipeline

A *pipeline* is a series of connected pipes through which a company sends oil or gas. One end of the pipeline originates at a production facility and the other end terminates at a facility on shore. Pumps or compressors inject the oil or gas into the pipeline. Often, producers tie in several small-diameter pipelines from several platforms to a single large-diameter pipeline that runs to shore (fig. 135). The smaller lines are *tie-in pipelines*. The large pipeline is a *trunk line*.

Building an offshore pipeline is a complex operation. It often involves the laying of several miles (kilometres) of large-diameter pipe on the seafloor under adverse conditions. Further, a pipeline construction company may have to dig a trench to protect the pipe after they lay it.

A pipeline construction company lays these lines on the seafloor. Also, the company puts the lines into trenches if it is necessary to protect the pipe from damage by fishing activity or other marine work. Usually, the construction company coats all but the ends of the pipe with concrete. Concrete protects the pipe, helps prevent corrosion caused by seawater, and weights the pipe so that it will stay on bottom. The construction crew leaves the pipe ends free of concrete so they can weld the joints together. They then coat the welded joints to prevent them from corroding.

If the pipeline carries crude oil, the terminal on shore will probably consist in part of several large storage tanks. From these tanks, land pipelines send oil to refineries.

A gas pipeline generally terminates at a facility capable of processing the gas. Gas processing involves

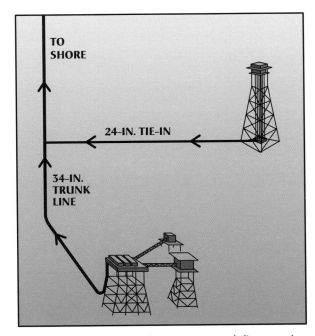

Figure 135. A tie-in pipeline joins a trunk line running to a shore facility.

recovering the heavier hydrocarbon components of natural gas, such as propane and butane. After the processing plant breaks down the gas into its components, it then pipelines the components to other plants for additional processing or to consumers.

In short, offshore pipeliners have to connect hundreds or thousands of individual joints of pipe on the water's surface. They then have to carefully lay the pipe on the seafloor without damaging or weakening it.

LAYING PIPELINE WITH A LAY BARGE

An offshore pipeline construction company can lay an offshore pipeline in several ways. Commonly, they use a special barge, a *lay barge*. It floats on the water's surface, or, if the water is rough, it floats in a semisubmerged state, much like a semisubmersible drilling rig.

Think of a lay barge as a floating offshore assembly line. Each step in the operation—welding, inspection, wrapping, laying—occurs at a station on the barge as a special machine conveys the pipe along the barge's length. Ideally, once the operation starts, it won't stop until the line is completely laid. Unfortunately, bad weather and unforeseen events can bring things to an abrupt halt and increase what are already very high costs.

A lay barge carries a large number of pipeline joints. It also carries the personnel and equipment necessary to weld the joints together, inspect them for integrity, apply a protective coating on the welded joints, and place the pipe on the seafloor (fig. 136).

Figure 136. This lay barge is at work in the North Sea.

Further, as the welding machines join each joint, the barge has to be able to move forward so that it lays the pipe in a steady and continuous motion with as few delays as possible. Since the lay barge usually cannot carry enough pipe to complete the job, supply boats shuttle pipe to the barge from a shore base.

The job of this floating assembly line is to lay pipe from a group of platforms offshore to a point on land. The first order of business is to get the barge and its cargo of pipe and equipment on the pipeline route and keep it there. At the same time, the barge has to move forward along the route as it puts the pipeline joints together and lays them on bottom. How do they do it? With anchors and anchor lines.

Anchoring

At least one anchor tethers the barge at each corner; usually, two more anchors secure it amidships on both its port and starboard sides. Sometimes, in addition to the anchors, tugboats help keep the barge on station. When the barge starts laying pipe, tension on the anchor lines holds it in position. As the welding machines join each joint of pipe and the pipe is ready to be submerged, the barge has to move forward. Personnel pull it forward by slacking off on the stern anchor lines and taking in slack on the bow anchor lines. Meanwhile, other personnel have to adjust the anchor lines amidships, and, if tugs are involved, their movements have to be coordinated, too. No wonder computers are being used more and more in pipe-laying operations.

Welding

As pulling on the bow anchor lines moves the barge forward, a heavy-duty crane hoists the joints of pipe onto a special conveyer belt—a *production ramp*. The production ramp conveys the joint to a *line-up station*. The line-up station aligns the end of the joint with the joint ahead of it. A special clamp holds the joints steady, and a welding machine automatically makes the first of several welds to start the connecting process. Usually, the lay barge makes several welds to ensure a good, strong joint. Welding machines at several stations on the barge make these welds as the barge moves forward.

Inspection and Wrapping

An inspector uses special equipment to examine each weld thoroughly after completion. It is difficult and expensive for a pipeline company to repair offshore pipelines after they have laid them on bottom. Also, leaks lose valuable hydrocarbons (not to mention creating possible pollution problems). At the barge's inspection station, special cameras X-ray each weld, and inspectors carefully examine the film to check for flaws. Should they find defects, they stop the operation until personnel correct the defects.

Once the weld passes inspection, the pipe moves to a wrapping station, where wrapping machines coat the welded joint (usually with concrete) to protect it. The coating keeps the joint from coming into direct contact with seawater and thus prevents corrosion.

Stingers and Tensioners

As the pipe moves off the stern of the barge and enters the water, it bends. Bending puts stress on the pipe. To help alleviate stress, the pipe slides over an extension of the production ramp—the *stinger.* The stinger supports the pipe on its way to the ocean floor (fig. 137). It is hinged to the barge's stern and personnel can adjust it up or down to decrease or increase the angle at which the pipe enters the water. The idea is to adjust the stinger's angle so that bend stresses are minimized.

To lay pipe in deep water, pipeline builders usually employ a *pipe tensioning system.* The tensioner presses several large wheels with tires on them against opposite sides of the pipe as it moves through them on its way into the water. Brakes on the wheels put tension on the pipe. The tensioner acts as a brake that supports the weight of the pipe before it comes to rest on the seafloor. You can think of tensioners as exerting an upward pull that offsets the downward pull caused by the pipe's weight. Keeping tension on the pipe helps prevent it from wrinkling and buckling under its own weight before it reaches bottom.

Figure 137. Protruding from the stern of this semisubmersible lay barge is the stinger, which supports the pipe as it enters the water.

Figure 138. Pipeline is wound onto the reel of this reel ship. *(Courtesy Santa Fe International Corp.)*

LAYING PIPELINE WITH A REEL VESSEL

If the pipeline is not too large in diameter—say, 16 inches (400 millimetres) or less—the pipeline construction company may use a reel ship or barge to lay it (fig. 138). The company constructs the pipeline at an onshore facility. There, personnel and machines weld, coat, and inspect the pipe, and wind it onto a huge reel. When reeling pipeline, the company does not use concrete to add weight. Concrete would break off as the pipe is reeled. If they need weight to prevent the pipe from floating, they use special thick-walled pipe.

A *reel ship* or *barge* handles the reeled-up pipeline. To lay the line, the vessel pays out the pipe off the reel at a steady rate onto the ocean floor. Unlike conventional lay barges, where companies spend most of their time laying the pipe, on reel ships or barges they spend little time doing the actual laying. Instead, the company spends most of the time winding pipe onto the reel and bringing it to the job site. This is advantageous because bad weather on land doesn't slow down construction nearly as much as it does offshore. The less time spent offshore, the less likely will bad weather hold up the operation.

LAYING PIPELINE BY THE BOTTOM-PULL METHOD

In shallow waters, pipeliners can sometimes lay pipe with the *bottom-pull method*. In this method, machines weld the pipe into a single, very long section or into several shorter sections at an onshore location. Towboats then pull the section or sections into the water. The pipe submerges and the vessel tows the pipe to the site. Once at the site, pipeline builders tie in the pipeline to the platform or other producing facility. They then head toward shore, following the already laid pipeline, and connect pipeline joints as necessary.

Sometimes, personnel spread pontoons or other flotation devices along the pipeline to give it buoyancy. Buoyancy allows the water to bear part of the pipe's weight, which eases the strain on the towing vessel.

Companies also use the bottom-pull method in some deepwater jobs, especially when they can float the pipe out to the site. Once again, when bad weather limits the time pipeliners have to lay pipe, the less time they must spend offshore, the better. Pulling an already constructed pipeline out to the site is much faster than constructing it on the site.

A PRIMER OF OFFSHORE OPERATIONS

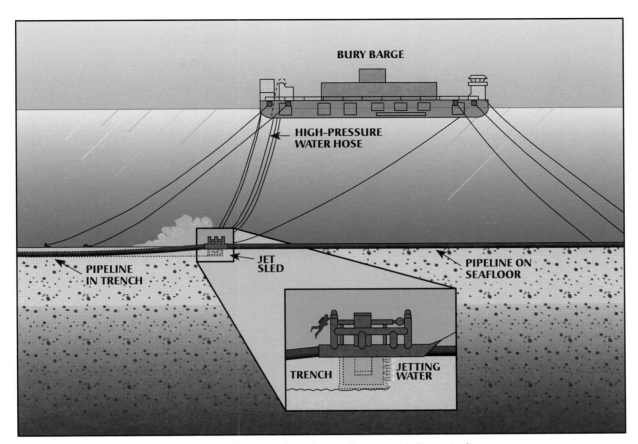

Figure 139. By using high-pressure jets of water, a bury barge digs a trench for a pipeline.

BURYING PIPELINE WITH A BURY BARGE

Regardless of how companies construct the pipeline, they may also have to bury it beneath the seafloor, especially in areas where marine activity may damage it. One way companies bury pipe is with a *bury barge* (fig. 139). Like a lay barge, a bury barge moves itself forward over the already laid pipe by means of anchors and lines. The bury barge lowers a *jet sled* over the pipeline. The sled straddles the pipe, and, as the barge pulls it over the pipe, high-pressure jets of water remove soil from beneath the pipe. The pipe then sags naturally into the jetted-out trench.

Transportation by Tanker

Sometimes, it is not feasible for a company to build a pipeline to carry oil to shore. For example, if an area's production is not very large, the expense of building a pipeline could be more than the income received from the sale of oil. Or, building a pipeline at a particular site could be physically impossible. Whatever the reason, when companies cannot pipeline oil to shore, they often use oil *tanker ships* instead.

In the case of an offshore gas field, the operator usually must build a pipeline because no other practical way exists to transport gas. In a few cases, however, a company may build an offshore gas-liquefaction facility—where the facility cools natural gas to an extremely low temperature and liquefies it. The company can then transport the liquid gas to shore via tankers specially designed to handle liquefied natural gas (LNG).

On the other hand, a company can produce oil from a platform or subsea wells and pipe it to a special mooring system. A tanker ship can tie up to the system, take on oil, and then transport the oil to an onshore terminal for unloading.

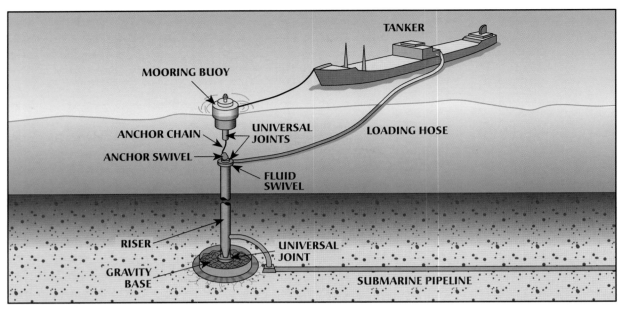

Figure 140. A single point buoy mooring system loads oil into a tanker.

The most common type of mooring system is *single-point buoy mooring* (*SPBM*) (fig. 140). An SPBM system is an installation in the sea where a tanker can take on a crude oil cargo. It is called a single-point system because it provides only one point to which the tanker ties up, or moors. Oil to the SPBM system comes from wells on a platform or from subsea wells.

One type of system consists of a gravity base, a riser, an anchor chain, and a buoy to which the tanker moors. The riser is a tube or a bundle of tubes through which oil rises to the loading hose. The hose usually has a ballast-and-buoyancy device to allow personnel to adjust the hose angle for efficient loading. The ballast-and-buoyancy device

also stabilizes the hose to prevent it from moving around too much in heavy seas.

Oil flows into the bottom of the riser through well flow lines. A loading hose near the top of the riser conveys oil into the tanker's cargo compartments. The riser has universal joints at the bottom and top of the riser. Also, the base of the anchor chain has a swivel. The universal joints and swivel allow the tanker to take on oil even as it rotates around the system under the influence of wind and currents.

The gravity base holds the whole SPBM system stationary on the seafloor. The base is so heavy that the force of gravity alone holds it in place. Personnel do not need to pin or pile it to the seafloor.

Summary

Moving oil and gas to shore is an essential step in offshore operations. Often, the oil company injects oil and gas into a pipeline. They lay the pipeline from a platform or group of platforms to an onshore facility. In areas where a company cannot lay a pipeline, they transport the oil to shore by using tankers.

A lay barge usually lays a pipeline. A lay barge is a large floating vessel that carries the people and equipment needed to construct and lay the pipeline on the seafloor. Personnel and machinery on the barge weld, inspect, wrap, and place the pipeline on bottom. Six anchors, and sometimes tugboats, keep the barge on station and pull it along the route as the barge lays the pipe.

Sometimes, a company constructs the pipeline on land, spools it onto a large reel, mounts the reel on a barge or ship, and carries the whole reeled pipeline on the reel vessel to the site. The reel vessel simply pays out the pipe onto the ocean bottom.

Unreeling pipe considerably shortens the time spent offshore in laying the pipe. This procedure can be an advantage when weather conditions on the route are bad enough to cause delays in conventional construction techniques.

The bottom-pull method of laying pipe also shortens the time needed to lay pipe on the route. In this method, the pipeline builder connects the pipe joints at a shore facility, and tows the long section of pipe to the site. If pipeline builders use flotation devices, which allow the water to take up some of the weight of the pipe, then they can use the bottom-pull method in deep and shallow waters.

If marine activities might damage the pipeline, the company can bury it using a bury barge. The bury barge pulls a special jet sled over the already laid pipeline. The jet sled uses high-pressure jets of seawater to scour out the soil from beneath the pipeline. As the sled digs the trench, the pipeline falls naturally into the trench.

Where companies cannot lay pipelines, they may use crude oil tankers to move oil to shore. Frequently, oil from the wells flows into a single-point buoy mooring system. The tanker ties up to the system, takes on oil, and transports it to shore.

REVIEW

Offshore operations cover a broad range of activities that include exploration, exploration drilling, development drilling, production, workover, and transportation.

Offshore exploration includes the use of magnetic, gravity, and seismic surveys. These survey methods yield evidence of the presence of hydrocarbons in formations that lie below the oceans and seas of the world. Should exploration techniques reveal the likelihood of hydrocarbons at a certain site, then drilling a well becomes necessary. Often, operators use a mobile offshore drilling unit—a MODU—to drill the well. Because companies can evaluate a well after they drill it, they can find out for sure whether oil and gas reside in a potential reservoir. Such evaluation techniques as logging, drill stem testing, and coring can confirm that the exploration well has penetrated a hydrocarbon-bearing formation.

If the wildcat well shows what appear to be ample quantities of oil and gas, a company may drill appraisal wells. Appraisal wells further confirm that the reservoir holds enough hydrocarbons to justify the enormous expense of developing the reservoir.

If commercial amounts of oil or gas do exist in a reservoir, the company must develop the reservoir. It may move a self-contained platform to the site and erect the platform. The platform may be a rigid steel or concrete structure, or it may be a compliant steel structure if the water is very deep. Usually, the operator drills several development wells from the platform. On the other hand, the operator may place a subsea template on the seafloor, and drill the wells from a MODU. Whether from a platform or on a template, the operator drills and completes several directional wells to produce hydrocarbons from the reservoir.

Producers may drill satellite wells where they cannot get to the reservoir from a platform or a template. Satellite wells and wells drilled on a template are often subsea completions; that is, producers install the Christmas trees on or below the seafloor. Wells that operators complete on platforms are usually surface completions. That is, they install the Christmas trees on a platform deck above the waterline.

The production phase of offshore operations includes the installation of special pieces of equipment to separate and treat crude oil, natural gas, and water. In most cases, operators install the equipment on a platform—usually the same platform they drilled the wells on. Sometimes, however, they install the equipment on a special processing and storage ship. They most often use such ships with subsea completions.

Operating companies may have to put wells produced from platforms and from special ships on artificial lift. They will most likely use gas lift, which allows wells to produce after natural drive depletes. In some cases, applying additional recovery techniques to extract still more hydrocarbons from an aged and dying reservoir may be possible.

Once a company completes and puts development wells on production, it will need to service and repair the wells to keep them producing at the desired rate. In many instances, operators can use wireline and pumpdown repair methods, but in other instances, they may have to move in a semisubmersible or jackup well service and workover unit.

Finally, a company may build a pipeline to transport the oil and gas to refineries or plants for the extraction of useful products. Or perhaps tanker ships will shuttle oil to shore. If so, the operator will install special mooring systems.

Offshore operations are occurring and will continue to occur all over the world. For it is there, beneath the waves, that the petroleum industry will find most of the world's future hydrocarbons.

GLOSSARY

A

annular space *n*: 1. the space surrounding a cylindrical object within a cylinder. 2. the space around a pipe in a wellbore, the outer wall of which may be the wall of either the borehole or the casing; sometimes termed the annulus.

annulus *n*: see *annular space*.

anomaly *n*: a deviation from the norm. In geology, the term indicates an abnormality such as a fault or dome in a sedimentary bed.

anticlinal trap *n*: a hydrocarbon trap in which petroleum accumulates in the top of an anticline. See *anticline*.

anticline *n*: an arched, inverted-trough configuration of folded and stratified rock layers. Compare *syncline*.

appraisal well *n*: a well drilled to confirm and evaluate the presence of hydrocarbons in a reservoir that has been found by a wildcat well.

arctic submersible rig *n*: a mobile submersible drilling structure used in arctic areas. The rig is towed onto the drilling site and submerged during periods when the water is free of ice. All equipment below the waterline is surrounded by a caisson to protect it from damage by moving ice. The drilling deck has no square corners so that moving ice can better flow around it. See *submersible drilling rig*.

artificial gravel island *n*: an artificial island sometimes used in shallow Arctic waters as a base on which drilling and production equipment is erected.

artificial lift *n*: any method used to raise oil to the surface through a well after reservoir pressure has declined to the point at which the well no longer produces by means of natural energy. Sucker rod pumps, gas lift, hydraulic pumps, and submersible electric pumps are the most common means of artificial lift.

B

ballast *n*: 1. for ships, water taken on board into specific tanks to permit proper angle of repose of the vessel in the water, and to assure structural stability. 2. for mobile offshore drilling rigs, weight added to make the rig more seaworthy, increase its draft, or submerge it to the seafloor. Seawater is usually used for ballast, but sometimes concrete or iron is also used to lower the rig's center of gravity permanently.

ballast-control specialist *n*: on a semisubmersible drilling rig, the person responsible for maintaining the rig's stability under all weather and load conditions.

barge *n*: 1. a flat-decked, shallow-draft vessel, usually towed by a boat. Barges are not self-propelled. They are used to transport oil or products on rivers, lakes, and inland waterways. Also, a complete drilling rig may be assembled on a barge and the vessel used for drilling wells in lakes and in inland waters and marshes. Similarly, well service and workover equipment can be mounted on a barge. 2. an offshore drilling vessel built in the shape of a ship. Unlike a ship, however, it is not self-propelled. Also called a drill barge.

basement rock *n*: igneous or metamorphic rock, which seldom contains petroleum. Ordinarily, it lies below sedimentary rock. When it is encountered in drilling, the well is usually abandoned.

bent sub *n*: a short cylindrical device installed in the drill stem between the bottommost drill collar and a downhole motor. Its purpose is to deflect the downhole motor off vertical to drill a directional hole. See *drill stem*.

bit *n*: the cutting or boring element used in drilling oil and gas wells. The bit consists of a cutting element and a circulating element. The circulating element permits the passage of drilling fluid and utilizes the hydraulic force of the fluid stream to improve drilling rates. In rotary drilling, several drill collars are joined to the bottom end of the drill pipe column, and the bit is attached to the end of the string of drill collars. Most bits used in rotary drilling are roller cone bits, but diamond bits are also used extensively.

block *n*: any assembly of pulleys on a common framework; in mechanics, one or more pulleys, or sheaves, mounted to rotate on a common axis. The crown block is an assembly of sheaves mounted on beams at the top of the derrick or mast. The drilling line is reeved over the sheaves of the crown block alternately with the sheaves of the traveling block, which is hoisted and lowered in the derrick or mast by the drilling line. When elevators are attached to a hook on the traveling block and drill pipe latched in the elevators, the pipe can be raised or lowered. See *crown block*, *traveling block*.

blowout preventer (BOP) *n*: one of several valves installed at the wellhead to prevent the escape of pressure either in the annular space between the casing and drill pipe or in open hole (i.e., hole with no drill pipe) during drilling or completion operations. Blowout preventers on land rigs are located beneath the rig at the land's surface; on jackup or platform rigs, at the water's surface; and on floating offshore rigs, on the seafloor.

blowout preventer stack (BOP stack) *n*: the assembly of well-control equipment including preventers, spools, valves, and nipples connected to the top of the wellhead.

BOP *abbr*: blowout preventer.

bottle-type semisubmersible rig *n*: see *semisubmersible drilling rig*.

bottle-type submersible rig *n*: a mobile submersible drilling structure constructed of several steel cylinders, or bottles. When the bottles are flooded, the rig submerges and rests on bottom; when water is removed from the bottles, the rig floats. The latest designs of this type of rig drill in water depths up to 100 feet (30.5 metres). See *submersible drilling rig*.

bottom-pull method *n*: an offshore pipeline construction technique in which the pipe string remains below the surface while it is towed to its final location.

bottom-supported offshore drilling rig *n*: a type of mobile offshore drilling unit (MODU) that has a part of its structure in contact with the seafloor when it is on site and drilling a well. The remainder of the rig is supported above the water. The rig can float, however, allowing it to be moved from one drill site to another. Bottom-supported units include submersible rigs and jackup rigs. See *mobile offshore drilling unit*.

break out *v*: 1. to unscrew one section of pipe from another section, especially drill pipe while it is being withdrawn from the wellbore. During this operation, the tongs are used to start the unscrewing operation. 2. to separate, as gas from a liquid or water from an emulsion.

bright spot *n*: a seismic phenomenon that shows up on a seismic, or record, section as a sound reflection that is much stronger than usual. A bright spot sometimes directly indicates natural gas in a trap.

bury barge *n*: a vessel used to bury pipeline beneath the seafloor. The barge moves forward by means of anchors. A jet sled is lowered over the pipeline, and as the barge pulls it over the pipe, high-pressure jets of water remove soil from beneath the pipe, allowing the pipe to fall into the jetted-out trench.

C

caisson *n*: 1. one of several columns made of steel or concrete that serve as the foundation for a rigid offshore platform rig, such as the concrete gravity platform rig. 2. a steel or concrete chamber that surrounds equipment below the waterline of an Arctic submersible rig, thereby protecting the equipment from damage by moving ice.

caisson-type platform rig *n*: a rigid offshore drilling platform that stands on steel caissons and is used to drill development wells. The caissons are firmly affixed to the seafloor, and the drilling and production decks are laid on top of them. The platform is used in certain Arctic waters, where the caissons are needed to protect equipment from moving ice. See *platform rig*.

cantilever *n*: a beam or beams that project outward from a structure and are supported only at one end.

cantilevered jackup *n*: a jackup drilling unit in which the drilling rig is mounted on two cantilevers that extend outward from the barge hull of the unit. The cantilevers are supported only at the barge end. Compare *keyway*.

carbonate *n*: 1. a salt of carbonic acid. 2. a compound containing the carbonate radical (CO_3). 3. a carbonate rock.

carbon dioxide *n*: a colorless, odorless gaseous compound of carbon and oxygen, CO_2. A product of combustion and a filler for fire extinguishers, this heavier-than-air gas can collect in low-lying areas, where it may displace oxygen and present the hazard of anoxia.

casing *n*: steel pipe placed in an oil or gas well as drilling progresses to prevent the wall of the hole from caving in, to prevent seepage of fluids, and to provide a means of extracting petroleum if the well is productive.

caving *n*: collapsing of the walls of the wellbore. Also called sloughing.

cement *n*: a powder, consisting of alumina, silica, lime, and other substances, that hardens when mixed with water. Extensively used in the oil industry to bond casing to the walls of the wellbore.

choke line *n*: a pipe attached to the blowout preventer stack out of which kick fluids and mud can be pumped to the choke manifold when a blowout preventer is closed in on a kick.

Christmas tree *n*: the control valves, pressure gauges, and chokes assembled at the top of a well to control the flow of oil and gas after the well has been drilled and completed.

circulating components *n pl*: the equipment included in the drilling fluid circulating system of a rotary rig. Basically, the components consist of the mud pump, rotary hose, swivel, drill stem, bit, and mud return line.

circulating system *n*: responsible for getting drilling mud down the hole through drill pipe and drill collars, jetting mud through the bit so it can pick up cuttings, and carrying cuttings to the surface.

clastics *n pl*: 1. sediments formed by the breakdown of large rock masses by climatological processes, physical or chemical. 2. the rocks formed from these sediments.

clump weights *n pl*: special segmented weights attached to the guy wires of a guyed compliant platform that keep the guy wires taut as the platform jacket moves with the waves and current of the water. As tension on a guy wire is lessened, one or more segments of the weight drop to the seafloor and reestablish tension on the wire. As tension increases on a guy wire, the tension pulls one or more segments off the seafloor so that the wire is not overstressed.

column-stabilized semisubmersible rig *n*: see *semisubmersible drilling rig*.

combination drive *n*: a combination of two or more natural energies that work together in a reservoir to force fluids into a wellbore. Possible combinations include gas-cap and water drive, solution gas and water drive, and gas-cap drive and gravity drainage.

combination trap *n*: a combination of two or more geologic structures (formations), such as an anticline and a fault, that are arranged such that hydrocarbons can accumulate.

company man *n*: see *company representative*.

company representative *n*: an employee of an operating company whose job is to represent the company's interests at the drilling location.

complete a well *v*: to finish work on a well and bring it to productive status. See *well completion*.

compliant platform rig *n*: an offshore platform that is designed to flex with wind and waves. Two types are the guyed-tower platform rig and the compliant piled tower.

concrete gravity rigid platform rig *n*: a rigid offshore drilling platform built of steel-reinforced concrete and used to drill development wells. The platform is floated to the drilling site in a vertical position, and at the site tall caissons that serve as the foundation of the platform are flooded so that the platform submerges and comes to rest on bottom. Because of the enormous weight of the platform, the force of gravity alone keeps it in place. See *platform rig*.

concrete island drilling system (CIDS) *n*: a submersible drilling unit whose equipment below the waterline is surrounded by concrete walls to protect the equipment from moving ice damage.

conductor casing *n*: generally, the first string of casing in a well. It may be lowered into a hole drilled into the formations near the surface and cemented in place; or it may be driven into the ground by a special pile driver (in such cases, it is sometimes called drive pipe); or it may be jetted into place in offshore locations. Its purpose is to prevent the soft formations near the surface from caving in and to conduct drilling mud from the bottom of the hole to the surface when drilling starts. Also called conductor pipe, drive pipe.

conductor pipe *n*: see *conductor casing*.

conical drilling unit (CDU) *n*: a submersible drilling unit designed specifically for arctic waters. Drilling takes place in the summer during a brief period of ice-free water. It withstands tremendous forces of pack ice that surrounds it throughout the year.

controlled directional drilling *n*: see *directional drilling*.

core n: a cylindrical sample taken from a formation for geological analysis. Usually a conventional core barrel is substituted for the bit and procures a sample as it penetrates the formation. *v*: to obtain a formation sample for analysis.

core barrel *n*: a tubular device, usually from 10 to 60 feet (3 to 18 metres) long, run at the bottom of the drill pipe in place of a bit and used to cut a core sample.

core bit *n*: a bit that does not drill out the center portion of the hole, but allows this center portion (the core) to pass through the round opening in the center of the bit and into the core barrel.

coring *n*: the process of cutting a vertical, cylindrical sample of the formations encountered as an oilwell is drilled. The purpose of coring is to obtain rock samples, or cores, in such a manner that the rock retains the same properties that it had before it was removed from the formation.

crane operator *n*: usually, the roustabout foreman.

crown block *n*: an assembly of sheaves mounted on beams at the top of the derrick or mast over which the drilling line is reeved. See *block*.

D

dehydrate *v*: to remove water from a substance. Dehydration of crude oil is normally accomplished by treating with emulsion breakers. The water vapor in natural gas must be removed to meet pipeline requirements; a typical maximum allowable water vapor content is 7 pounds per million cubic feet (3.2 kilograms per million cubic metres) per day.

density *n*: the mass or weight of a substance per unit volume. For instance, the density of a drilling mud may be 10 pounds per gallon (ppg), 74.8 pounds per cubic foot (lb/ft³), or 1,198.2 kilograms per cubic metre (kg/m³). Specific gravity, relative density, and API gravity are other units of density.

derrick *n*: a large load-bearing structure, usually of bolted construction. In drilling, the standard derrick has four legs standing at the corners of the substructure and reaching to the crown block. The substructure is an assembly of heavy beams used to elevate the derrick and provide space to install blowout preventers, casingheads, and so forth. Because the standard derrick must be assembled piece by piece, it has largely been replaced by the mast, which can be lowered and raised without disassembly. Compare *mast*.

derrickhand *n*: the crew member who handles the upper end of the drill string as it is being hoisted out of or lowered into the hole. Also responsible for the circulating machinery and the conditioning of the drilling fluid.

desander *n*: a centrifugal device for removing sand from drilling fluid to prevent abrasion of the pumps. It may be operated mechanically or by a fast-moving stream of fluid inside a special cone-shaped vessel, in which case it is sometimes called a hydrocyclone. Compare *desilter*.

desilter *n*: a centrifugal device for removing very fine particles, or silt, from drilling fluid to keep the amount of solids in the fluid at the lowest possible point. Usually, the lower the solids content of mud, the faster is the rate of penetration. The desilter works on the same principle as a desander. Compare *desander*.

development drilling *n*: drilling that occurs after the initial discovery of hydrocarbons in a reservoir. Usually, several wells are required to adequately develop a reservoir.

development well *n*: 1. a well drilled in proven territory in a field to complete a pattern of production. 2. an exploitation well.

diesel electric rig *n*: see *electric rig*.

diesel engine *n*: a high-compression, internal-combustion engine used extensively for powering drilling rigs. In a diesel engine, air is drawn into the cylinders and compressed to very high pressures; ignition occurs as fuel is injected into the compressed and heated air. Combustion takes place within the cylinder above the piston, and expansion of the combustion products imparts power to the piston.

directional drilling *n*: intentional deviation of a wellbore from the vertical. Although wellbores are normally drilled vertically, it is sometimes necessary or advantageous to drill at an angle from the vertical. Controlled directional drilling makes it possible to reach subsurface areas laterally remote from the point where the bit enters the earth. It often involves the use of deflection tools.

dissolved-gas drive *n*: a source of natural reservoir energy in which the dissolved gas coming out of the oil expands to force the oil into the wellbore. Also called solution-gas drive. See *reservoir drive mechanism*.

diving bell *n*: a cylindrical or spherical compartment used to transport a diver or dive team to and from an underwater work site.

dome *n*: a geological structure resembling an inverted bowl, i.e., a short anticline that dips or plunges on all sides.

dome plug trap *n*: a reservoir formation in which fluid or plastic masses of rock material originated at unknown depths and pierced or lifted the overlying sedimentary strata.

downhole motor *n*: a drilling tool made up in the drill string directly above the bit. It causes the bit to turn while the drill string remains fixed. It is used most often as a deflection tool in directional drilling, where it is made up between the bit and a bent sub (or, sometimes, the housing of the motor itself is bent). Two principal types of downhole motor are the positive-displacement motor and the downhole turbine motor. Also called mud motor.

draft *n*: the vertical distance between the bottom of a vessel floating in water and the waterline.

drawworks *n*: the hoisting mechanism on a drilling rig. It is essentially a large winch that spools off or takes in the drilling line and thus raises or lowers the drill stem and bit.

drill collar *n*: a heavy, thick-walled tube, usually steel, placed between the drill pipe and the bit in the drill stem to provide a pendulum effect to the drill stem and weight to the bit.

driller *n*: the employee directly in charge of a drilling or workover rig and crew. The driller's main duty is operation of the drilling and hoisting equipment, but also responsible for downhole condition of the well, operation of downhole tools, and pipe measurements.

drilling contractor *n*: an individual or group of individuals who own a drilling rig or mast and contract their services for drilling wells.

drilling draft *n*: the depth at which a drilling rig rests below the water's surface.

drilling fluid *n*: circulating fluid, one function of which is to force cuttings out of the wellbore and to the surface. Other functions are to cool the bit and to counteract downhole formation pressure. While a mixture of barite, clay, water, and chemical additives is the most common drilling fluid, wells can also be drilled by using air, gas, water, or oil-base mud as the drilling fluid. See *mud*.

drilling line *n*: a wire rope used to support the drilling tools. Also called the rotary line.

drilling mud *n*: a specially compounded liquid circulated through the wellbore during rotary drilling operations. See *drilling fluid, mud*.

drilling platform rig *n*: see *platform rig*.

drilling slot *n*: see *keyway*.

drilling template *n*: see *temporary guide base*.

drilling tender *n*: a combination barge-and-platform design that can be towed to a new location after a well is drilled.

drill pipe *n*: heavy seamless tubing used to rotate the bit and circulate the drilling fluid. Joints of pipe approximately 30 feet (9 metres) long are coupled together by means of tool joints.

drill ship *n*: a self-propelled floating offshore drilling unit that is a ship constructed to permit a well to be drilled from it. While not as stable as semisubmersibles, drill ships are capable of drilling exploratory wells in deep, remote waters. They may have a ship hull, a catamaran hull, or a trimaran hull. See *floating offshore drilling rig*.

drill stem *n*: all members in the assembly used for rotary drilling from the swivel to the bit, including the kelly, drill pipe and tool joints, drill collars, stabilizers, and various specialty items. Compare *drill string*.

drill stem test (DST) *n*: the conventional method of formation testing. The basic drill stem test tool consists of a packer or packers, valves or ports that may be opened and closed from the surface, and two or more pressure-recording devices. The tool is lowered on the drill string to the zone to be tested. The packer or packers are set to isolate the zone from the drilling fluid column. The valves or ports are then opened to allow for formation flow while the recorders chart static pressures. A sampling chamber traps clean formation fluids at the end of the test. Analysis of the pressure charts is an important part of formation testing.

drill string *n*: the column, or string, of drill pipe with attached tool joints that transmits fluid and rotational power from the kelly to the drill collars and bit. Often, especially in the oil patch, the term is loosely applied to both drill pipe and drill collars. Compare *drill stem*.

drive pipe *n*: see *conductor casing*.

dry glycol *n*: glycol that has not absorbed water.

dynamic positioning system *n*: the means by which a floating drilling vessel maintains its position over a well location by varying the power on a series of propulsion units mounted along the hulls of the vessel. Propulsion system power is directed by a system of telemetry signals from beacons on the sea floor, satellite information, or the angular movements of a cable.

E

electric rig *n*: a drilling rig on which the energy from the power source—usually several diesel engines—is changed to electricity by generators mounted on the engines. The electrical power is then distributed through electrical conductors to electric motors. The motors power the various rig components. Compare *mechanical rig*.

electrostatic treater *n*: a vessel that receives an emulsion and resolves the emulsion to oil, water, and usually gas, by using heat, chemicals, and a high-voltage electric field. This field, produced by grids placed perpendicular to the flow of fluids in the treater, aids in breaking the emulsion. Also called an electrochemical treater. See *emulsion treating*.

emulsified water *n*: water so thoroughly combined with oil that special treating methods must be applied to separate it from the oil. Compare *free water*.

emulsion *n*: a mixture in which one liquid, termed the dispersed phase, is uniformly distributed (usually as minute globules) in another liquid, called the continuous phase or dispersion medium. In an oil-in-water emulsion, the oil is the dispersed phase and the water the dispersion medium; in a water-in-oil emulsion, the reverse holds. A typical product of oilwells, water-in-oil emulsion is also used as a drilling fluid.

emulsion treating *n*: the process of breaking down emulsions to separate oil from water or other contaminants. Treating plants may use a single process or a combination of processes to effect demulsification, depending on what emulsion is being treated.

escape capsule *n*: an emergency water-borne craft into which offshore rig personnel can enter and use to escape from the rig.

exploitation *n*: the development of a reservoir to extract its oil.

exploitation well *n*: a well drilled to permit more effective extraction of oil from a reservoir. Sometimes called a development well. See *development well*.

exploration *n*: the search for reservoirs of oil and gas, including aerial and geophysical surveys, geological studies, core testing, and drilling of wildcats.

exploration well *n*: see *wildcat*. Also called exploratory well.

F

fault *n*: a break in the earth's crust, or subsurface strata. Often strata on one side of the fault line have been displaced (upward, downward, or laterally) relative to their original positions.

fire tube *n*: a pipe, or set of pipes, within a tank through which steam or hot gases are passed to warm a liquid or gas in the tank. See *steam coil*.

fixed platform *n*: a structure made of steel or concrete, firmly fixed to the bottom of the body of water in which it rests.

flexible joint (flex joint) *n*: a device that provides a flexible connection between the riser pipe and the subsea blowout preventers. By accommodating lateral movement of a mobile offshore drilling rig, flex joints help to prevent a buildup of abnormal bending load pressure.

floater *n*: see *floating offshore drilling rig*.

floating offshore drilling rig *n*: a type of mobile offshore drilling unit that floats and is not secured to the seafloor (except with anchors) when it is in the drilling mode. Floating units include inland barge rigs, drill ships and ship-shaped barges, and semisubmersibles. See *mobile offshore drilling unit*.

floorhand *n*: see *rotary helper*.

flow line *n*: the surface pipe through which oil travels from a well to processing equipment or to storage.

fold *n*: a flexure of rock strata (e.g., an arch or a trough) produced by horizontal compression in the earth's crust. See *anticline, syncline*.

foundation-pile casing *n*: the first casing or conductor string (generally with a diameter of 30 to 36 inches (76 to 91 centimetres) set when drilling a well from a floating offshore drilling rig. It prevents sloughing of the ocean-floor formations and is a structural support for the permanent guide base and the blowout preventers.

foundation-pile housing *n*: equipment attached to the first joint of casing to go into a well drilled from a floating unit. The housing supports the foundation-pile casing when it lands in the temporary guide base. See *foundation pile casing, temporary guide base*.

free water *n*: 1. water produced with oil. It usually settles out within five minutes when the well fluids become stationary in a settling space within a vessel. Compare *emulsified water*. 2. the measured volume of water that is present in a container and that is not in suspension in the contained liquid at observed temperature.

free-water knockout (FWKO) *n*: a vertical or horizontal vessel into which oil or emulsion is run to allow any water not emulsified with the oil (free water) to drop out.

G

gas-cap drive *n*: drive energy supplied naturally (as a reservoir is produced) by the expansion of the gas cap. In such a drive, the gas cap expands to force oil into the well and to the surface. See *reservoir drive mechanism*.

gas lift *n*: the process of raising or lifting fluid from a well by injecting gas down the well through tubing or through the tubing-casing annulus. Injected gas aerates the fluid to make it exert less pressure than the formation does; consequently, the higher formation pressure forces the fluid out of the wellbore. Gas may be injected continuously or intermittently, depending on the producing characteristics of the well and the arrangement of the gas-lift equipment.

gas-lift mandrel *n*: a device installed in the tubing string of a gas-lift well onto which or into which a gas-lift valve is fitted. There are two common types of mandrel. In the conventional gas-lift mandrel, the gas-lift valve is installed as the tubing is placed in the well. Thus, to replace or repair the valve, the tubing string must be pulled. In the side-pocket mandrel, however, the valve is installed and removed by wireline while the mandrel is still in the well, eliminating the need to pull the tubing to repair or replace the valve.

gas-lift valve *n*: a device installed on a gas-lift mandrel, which in turn is put on the tubing string of a gas-lift well. Tubing and casing pressures cause the valve to open and close, thus allowing gas to be injected into the fluid in the tubing to cause the fluid to rise to the surface.

geopositioning satellite (GPS) system *n*: a system of satellites orbiting the earth that transmits and receives information to a special receiver and transmitter on the earth's surface. When the surface unit is activated it signals the satellite, which pinpoints the surface unit's location.

glycol *n*: a group of compounds used to dehydrate gaseous or liquid hydrocarbons or to inhibit the formation of hydrates. Glycol is also used in engine radiators as an antifreeze. Commonly used glycols are ethylene glycol, diethylene glycol, and triethylene glycol.

glycol dehydrator *n*: a processing unit used to remove all or most of the water from gas. A glycol unit usually includes an absorber, in which the wet gas is put into contact with glycol to remove the water, and a reboiler, which heats the wet glycol to remove the water from it so that it can be recycled.

gravimeter *n*: a device for measuring and recording the density or specific gravity of a gas or liquid passing a point of measurement.

gravity survey *n*: an exploration method in which an instrument that measures the intensity of the earth's gravity is passed over the surface or through the water. In places where the instrument detects stronger- or weaker-than-normal gravity forces, a geologic structure containing hydrocarbons may exist.

guide frame *n*: in drilling from floaters, a device fitted around the end of the drill stem into which guidelines on the permanent guide base are threaded. The guide frame guides the bit and drill stem into the wellbore. See *guidelines*.

guidelines *n pl*: lines, usually four, attached to the temporary guide base and permanent guide base to help position equipment (such as blowout preventers) accurately on the seafloor when a well is drilled offshore from a floating vessel.

guyed-tower platform rig *n*: a compliant offshore drilling platform used to drill development wells. The foundation of the platform is a relatively lightweight jacket upon which all equipment is placed. A system of guy wires anchored by clump weights helps secure the jacket to the seafloor and allows it to move with wind and wave forces. See *platform rig*.

heater-treater *n*: a vessel that heats an emulsion and removes water and gas from the oil to raise it to a quality acceptable for a pipeline or other means of transport. A heater-treater is a combination of a heater, free-water knockout, and oil and gas separator.

heave *n*: the vertical motion of a ship or a floating offshore drilling rig.

heave compensator *n*: a device that moves with the heave of a floating offshore drilling rig to prevent the bit from being lifted off the bottom of the hole and then dropped back down (i.e., to maintain constant weight on the bit). It is used with devices such as bumper subs. See *motion compensator*.

helipad *n*: on most drill ships, a landing pad for helicopters.

highs *n pl*: in geology, subsurface formations that arch upward.

hoist *n*: 1. an arrangement of pulleys and wire rope or chain used for lifting heavy objects; a winch or similar device. 2. the drawworks. *v*: to raise or lift.

hoisting components *n pl*: drawworks, drilling line, and traveling and crown blocks. Auxiliary hoisting components include catheads, catshaft, and air hoist.

hoisting system *n*: the system on the rig that performs all the lifting on the rig, primarily the lifting and lowering of drill pipe out of and into the hole. It is composed of drilling line, traveling block, crown block, and drawworks. See also *hoisting components*.

hole opener *n*: a device, with teeth arranged on its outside circumference, used to enlarge the size of an existing borehole as it rotates.

hook *n*: a large, hook-shaped device from which the swivel is suspended. It is designed to carry maximum loads ranging from 100 to 650 tons (90 to 590 tonnes) and turns on bearings in its supporting housing. A strong spring within the assembly cushions the weight of a stand (90 feet, or about 27 metres) of drill pipe, thus permitting the pipe to be made up and broken out with less damage to the tool joint threads. Smaller hooks without the spring are used for handling tubing and sucker rods. See also *stand*, *swivel*.

horizontal drilling *n*: deviation of the borehole at least 80° from vertical so that the borehole penetrates a productive formation in a manner parallel to the formation.

hydrate *n*: a hydrocarbon and water compound that is formed under reduced temperature and pressure in gathering, compression, and transmission facilities for gas. Hydrates often accumulate in troublesome amounts and impede fluid flow. They resemble snow or ice. *v*: to enlarge by taking water on or in.

hydrocarbons *n pl*: organic compounds of hydrogen and carbon whose densities, boiling points, and freezing points increase as their molecular weights increase.

Although composed of only two elements, hydrocarbons exist in a variety of compounds because of the strong affinity of the carbon atom for other atoms and for itself. The smallest molecules of hydrocarbons are gaseous; the largest are solids. Petroleum is a mixture of many different hydrocarbons.

hydrocyclone *n*: a cone-shaped separator for separating various sizes of particles and liquid by centrifugal force. See *desander*, *desilter*.

hydrophone *n*: a device trailed in an array behind a boat in offshore seismic exploration that is used to detect sound reflections, convert them to electric current, and send them through a cable to recording equipment on the boat.

igneous rock *n*: a rock mass formed by the solidification of magma (molten rock) within the earth's crust or onto its surface. Granite is an igneous rock.

impermeable *adj*: preventing the passage of a fluid. A formation may be porous yet impermeable if there is an absence of connecting passages between the voids within it. See *permeability*.

impervious *adj*: see *impermeable*.

individual riser joints *n pl*: see *riser pipe*.

inland barge rig *n*: a floating offshore drilling structure consisting of a barge on which the drilling equipment is constructed. When moved from one location to another, the barge floats. When stationed on the drill site, the barge can be anchored in the floating mode or submerged to rest on the bottom. Typically, inland barge rigs are used to drill wells in marshes, shallow inland bays, and areas where the water covering the drill site is not too deep. Also called swamp barge. See *floating offshore drilling rig*.

intermediate casing string *n*: the string of casing set in a well after the surface casing but before production casing is set to keep the hole from caving and to seal off troublesome formations. In deep wells, one or more intermediate strings may be required. Sometimes called protection casing.

internal-combustion engine *n*: a heat engine in which the pressure necessary to produce motion of the mechanism results from the ignition or burning of a fuel-air mixture within the engine cylinder.

jacket *n*: 1. a tubular piece of steel in a tubing liner-type of sucker rod pump, inside of which is placed an accurately bored and honed liner. In this type of sucker rod pump, the pump plunger moves up and down within the liner and the liner is inside the jacket. 2. a platform jacket.

jackup *n*: see *jackup drilling rig*.

jackup drilling rig *n*: a mobile bottom-supported off-shore drilling structure with columnar or open-truss legs that support the deck and hull. When positioned over the drilling site, the bottoms of the legs rest on the seafloor. A jackup rig is towed or propelled to a location with its legs up. Once the legs are firmly positioned on the bottom, the deck and hull height are adjusted and leveled. Also called self-elevating drilling unit. See *bottom-supported offshore drilling rig*.

jet *n*: in a perforating gun using shaped charges, a highly penetrating, fast-moving stream of exploded particles that forms a hole in the casing, cement, and formation.

jet nozzle *n*: see *nozzle*.

jet sled *n*: in pipeline construction offshore, a pipe-straddling device fitted with nozzles on either side that is lowered by a bury barge. As water is pumped at high pressure through the nozzles, spoil from beneath the pipe is removed and pumped to one side of the trench. The line then sags naturally into position in the trench.

joint *n*: a single length (from 16 feet to 48 feet, or 4.88 metres to 14.63 metres) of drill pipe, drill collar, casing, or tubing that has threaded connections at both ends. Several joints screwed together constitute a stand of pipe.

K

kelly *n*: the heavy steel tubular device, four- or six-sided, suspended from the swivel through the rotary table and connected to the top joint of drill pipe to turn the drill stem as the rotary table turns. It has a bored passageway that permits fluid to be circulated into the drill stem and up the annulus, or vice versa.

kelly drive bushing *n*: a special device that, when fitted into the master bushing, transmits torque to the kelly and simultaneously permits vertical movement of the kelly to make hole. It may be shaped to fit the rotary opening or have pins for transmitting torque. Also called the drive bushing. See *kelly*.

keyway *n*: a slot in the edge of the barge hull of a jackup drilling unit over which the drilling rig is mounted and through which drilling tools are lowered and removed from the well being drilled. Compare *cantilevered jackup*.

kick *n*: an entry of water, gas, oil, or other formation fluid into the wellbore during drilling. It occurs because the pressure exerted by the column of drilling fluid is not great enough to overcome the pressure exerted by the fluids in the formation drilled. If prompt action is not taken to control the kick or kill the well, a blowout may occur.

kill line *n*: a pipe attached to the blowout preventer stack, into which mud or cement can be pumped to overcome the pressure of a kick. Sometimes used when normal kill procedures (circulating kill fluids down the drill stem) are not sufficient.

L

lay barge *n*: a barge used in the construction and placement of underwater pipelines. Joints of pipe are welded together and then lowered off the stern of the barge as it moves ahead.

line-up station *n*: on a pipeline lay barge, a place (station) on the barge where the end of one pipeline joint is aligned with the joint ahead of it.

LMRP *abbr*: lower marine riser package.

log *n*: a systematic recording of data, such as a driller's log, mud log, electrical well log, or radioactivity log. Many different logs are run in wells to obtain various characteristics of downhole formations.

lower marine riser package (LMRP) *n*: part of a blowout preventer; contains a wellhead connector and an annular-type blowout preventer. The LMRP is assembled with the blowout preventer and lowered to the seafloor on joints of marine riser pipe.

M

magnetic survey *n*: an exploration method in which an instrument that measures the intensity of the natural magnetic forces existing in the earth's subsurface is passed over the surface or through the water. The instrument can detect deviations in magnetic forces, and such deviations may indicate the existence of an underground hydrocarbon reservoir.

magnetometer *n*: an instrument used to measure the intensity and direction of a magnetic field, especially that of the earth.

make a trip *v*: to hoist the drill stem out of the wellbore to perform one of a number of operations, such as changing bits or taking a core, and so forth, and then to return the drill stem to the wellbore.

make up *v*: 1. to assemble and join parts to form a complete unit (e.g., to make up a string of casing). 2. to screw together two threaded pieces. 3. to mix or prepare (e.g., to make up a tank of mud). 4. to compensate for (e.g., to make up for lost time).

marine riser connector *n*: a fitting on top of the subsea blowout preventers to which the riser pipe is connected.

marine riser pipe *n*: see *riser pipe*.

marine riser system *n*: see *riser pipe*.

marine riser tensioner *n*: see *tensioner system*.

mast *n*: a portable derrick that is capable of being erected as a unit, as distinguished from a standard derrick, which cannot be raised to a working position as a unit. For transporting by land, the mast can be divided into two or more sections to avoid excessive length extending from truck beds on the highway. Compare *derrick*.

master bushing *n*: a device that fits into the rotary table to accommodate the slips and drive the kelly bushing so that the rotating motion of the rotary table can be transmitted to the kelly. Also called rotary bushing. See *kelly drive bushing*, *slips*.

mat-supported jackup rig *n*: a type of bottom support for the legs of a jackup, generally used in areas where the seafloor is very soft. Each leg of the jackup is connected to the mat, which is steel frame that is relatively wide. The mat, since it is wide, prevents the jackup's legs from penetrating the soft seafloor.

mechanical rig *n*: a drilling rig in which the source of power is one or more internal-combustion engines and in which the power is distributed to rig components through mechanical devices (such as chains, sprockets, clutches, and shafts). Also called a power rig. Compare *electric rig*.

metamorphic rock *n*: a rock derived from preexisting rocks by mineralogical, chemical, and structural alterations caused by processes within the earth's crust. Marble is a metamorphic rock.

miscible *adj*: 1. capable of being mixed. 2. capable of mixing in any ratio without separation of the two phases.

miscible flooding *n*: a method of secondary recovery of fluids from a reservoir by injection of fluids that will mix with the reservoir fluids.

mobile arctic caisson (MAC) *n*: a specially designed submersible offshore drilling rig that can withstand the force of moving ice.

mobile offshore drilling unit (MODU) *n*: a drilling rig that is used exclusively to drill offshore exploration and development wells and that floats on the surface of the water when being moved from one drill site to another. It may or may not float once drilling begins. Two basic types of mobile offshore drilling units are used to drill most offshore wildcat wells: bottom-supported drilling rigs and floating drilling rigs.

MODU *abbr*: mobile offshore drilling unit.

monkeyboard *n*: the derrickhand's working platform, which may be as high as 90 feet (27 metres) or higher in the derrick or mast. As pipe or tubing is run into or out of the hole, the derrickhand must handle the top end of the pipe (set it back in or take it out of the fingerboard). The monkeyboard provides a small platform to raise the derrickhand to the proper height for handling the top of the pipe.

moon pool *n*: a walled round hole or well in the hull of a drill ship (usually in the center) through which the drilling assembly and other assemblies pass while a well is being drilled, completed, or abandoned from the drill ship.

motion compensator *n*: any device (such as a bumper sub or heave compensator) that serves to maintain constant weight on the bit in spite of vertical motion of a floating offshore drilling rig.

mud *n*: the liquid, usually placed in steel tanks or pits on a rig, that is circulated through the wellbore during rotary drilling and workover operations. In addition to its function of bringing cuttings to the surface, drilling mud cools and lubricates the bit and drill stem, protects against blowouts by holding back subsurface pressures, and deposits a mud cake on the wall of the borehole to prevent loss of fluids to the formation. See *drilling fluid*.

mud pit *n*: originally, an open pit dug in the ground to hold drilling fluid or waste materials discarded after the treatment of drilling mud. For some drilling operations, mud pits are used for suction to the mud pumps, settling of mud sediments, and storage of reserve mud. Steel tanks are much more commonly used for these purposes now, but they are still sometimes referred to as pits; however, "mud tanks" is preferred.

mud pump *n*: a large, high-pressure reciprocating pump used to circulate the mud on a drilling rig. A typical mud pump is a single- or double-acting, two- or three-cylinder piston pump whose pistons travel in replaceable liners and are driven by a crankshaft actuated by an engine or a motor. Also called a slush pump.

mud return line *n*: a trough or pipe that is placed between the surface connections at the wellbore and the shale shaker and through which drilling mud flows on its return to the surface from the hole. Also called flow line.

mud tank *n*: see *mud pit*.

N

natural gas *n*: a highly compressible, highly expandable mixture of hydrocarbons with a low specific gravity and occurring naturally in gaseous form.

negative gravity anomaly *n*: on the record made by a gravity survey, an indication that less dense rocks lie under sedimentary rocks. For example, salt domes are less dense than the overlying sedimentary rocks, and usually show up as a negative gravity anomaly.

nipple up *v*: in drilling, to assemble the blowout preventer stack on the wellhead at the surface.

nonporous *adj*: containing no interstices; having no pores and therefore unable to hold fluids.

nozzle *n*: a passageway through jet bits that causes the drilling fluid to be ejected from the bit at high velocity. The jets of mud clean the bottom of the hole.

O

OCS *abbr*: Outer Continental Shelf.

oil and gas separator *n*: a piece of production equipment used to separate liquid components of the well stream from gaseous elements. Separators are either vertical or horizontal and either cylindrical or spherical in shape. Separation is accomplished principally by gravity, with the heavier liquids falling to the bottom and the gas rising to the top. A float valve or other liquid-level control regulates the level of oil in the bottom of the separator.

oil operator *n*: see *operator*.

operator *n*: the person or company, either proprietor or lessee, actually operating an oilwell or lease. Generally, the oil company that hires the drilling contractor.

organic theory *n*: an explanation of the origin of petroleum, which holds that the hydrogen and the carbon that make up petroleum come from land and sea plants and animals. Furthermore, the theory holds that more of this organic material comes from very tiny swamp and sea creatures than comes from larger land creatures.

Outer Continental Shelf (OCS) *n*: an offshore area in the United States that begins where state ownership of mineral rights ends and ends where international treaties dictate.

P

packer *n*: a piece of downhole equipment consisting of a sealing device, a holding or setting device, and an inside passage for fluids; used to block the flow of fluids through the annular space between the tubing and the wall of the wellbore by sealing off the space between them. It is usually made up in the tubing string some distance above the producing zone. A packing element expands to prevent fluid flow except through the inside bore of the packer and into the tubing. Packers are classified according to configuration, use, and method of setting and whether or not they are retrievable (i.e., whether they can be removed when necessary, or whether they must be milled or drilled out and thus destroyed).

perforate *v*: to pierce the casing wall and cement to provide holes through which formation fluids may enter or to provide holes in the casing so that materials may be introduced into the annulus between the casing and the wall of the borehole. Perforating is accomplished by lowering into the well a perforating gun, or perforator, that fires electrically detonated bullets or shaped charges.

permeability *n*: 1. a measure of the ease with which a fluid flows through the connecting pore spaces of rock or cement. The unit of measurement is the millidarcy. 2. fluid conductivity of a porous medium. 3. ability of a fluid to flow within the interconnected pore network of a porous medium.

permeable *adj*: allowing the passage of fluid. See *permeability*.

petroleum *n*: a substance occurring naturally in the earth and composed mainly of mixtures of chemical compounds of carbon and hydrogen, with or without other nonmetallic elements such as sulfur, oxygen, and nitrogen. The compounds that compose it may be in the gaseous, liquid, or solid state, depending on their nature and on temperature and pressure conditions.

pipeline *n*: a system of connected lengths of pipe, usually buried in the earth or laid on the seafloor, that is used for transporting petroleum and natural gas.

pipe tensioning system *n*: a braking system used on a lay barge to control the descent rate of the pipe. Tensioners also support the entire submerged weight of the pipe as it approaches the bottom.

pitch *n*: on offshore floating rigs, the up-and-down movement of the hull from the bow to the stern.

pit watcher *n*: assists the derrickhand in keeping tabs on the drilling mud and the circulating equipment.

platform rig *n*: an immobile offshore structure from which development wells are drilled and produced. Platform rigs may be built of steel or concrete and may be either rigid or compliant. Rigid platform rigs, which rest on the seafloor, are the caisson-type platform, the concrete gravity platform, and the steel-jacket platform. Compliant platform rigs, which are used in deeper waters and yield to water and wind movements, are the guyed-tower platform and the tension-leg platform.

plug *n*: any object or device that blocks a hole or passageway (such as a cement plug in a borehole).

pore *n*: an opening or space within a rock or mass of rocks, usually small and often filled with some fluid (water, oil, gas, or all three).

positive gravity anomaly *n*: on the record made by a gravity survey, an indication that dense rock may have intruded into overlying sedimentary rocks and uplifted them. Sometimes, hydrocarbons accumulate in traps formed by such intrusions.

posted barge submersible rig *n*: a mobile submersible drilling structure consisting of a barge hull that rests on bottom, steel posts that rise from the top of the barge hull, and a deck that is built on top of the posts, well above the waterline. It is used to drill wells in water no deeper than about 30–35 feet (9–10.7 metres). Most posted barge submersibles work in inland gulfs and bays. See *submersible drilling rig*.

pressure maintenance *n*: the use of waterflooding or natural gas recycling during primary recovery to provide additional formation pressure and displacement energy that can supplement and conserve natural reservoir drives. Although commonly begun during primary production, pressure maintenance methods are often considered to be a form of enhanced oil recovery.

prime mover *n*: an internal-combustion engine or a turbine that is the source of power for driving a machine or machines.

production *n*: 1. the phase of the petroleum industry that deals with bringing the well fluids to the surface and separating them and with storing, gauging, and otherwise preparing the product for the pipeline. 2. the amount of oil or gas produced in a given period.

production casing *n*: the last string of casing set in a well, inside of which is usually suspended a tubing string.

production platform *n*: see *platform rig*.

production ramp *n*: on a pipeline lay barge, a special conveyor on which pipeline joints are placed prior to their being joined and lowered to the seafloor.

production riser *n*: pipe and special fittings used to connect a subsea wellhead to a floating vessel, such as a tanker.

pumpdown tools *n pl*: for servicing subsea completions, special tools that can be pumped down the flow line and into the tubing string to do a specific job.

R

radio triangulation *n*: a method of locating a specific point of an object (such as a boat or airplane) by receiving radio transmissions from at least two different fixed stations, usually located on land. Both the shore-based stations and the portable station on the boat or airplane transmit and receive signals. The time it takes for the signal to travel between the portable station and each of the land stations indicates the precise distance between the portable station and the two permanent stations. With the two distances, the portable object's location can be accurately determined.

reboiler *n*: in a glycol dehydration unit, a device that heats wet glycol (glycol saturated with water picked up from the wet gas) to drive off the water. Since glycol boils at a higher temperature than water, the reboiler boils off the water and leaves the dry glycol behind.

record section *n*: a cross section of the earth generated by computer from tapes that have recorded the sound vibrations reflected during seismic exploration. Expert interpretation can reveal what may be a trap for petroleum. Also called seismic section.

reel vessel *n*: a ship or barge specially designed to handle pipeline that is wound onto a large reel. To lay the pipeline, the vessel pays out the pipe off the reel at a steady rate onto the ocean floor. The pipeline has been constructed at an onshore facility where it has been welded, coated, inspected, and wound onto the reel.

reeve *v*: to pass (as a rope) through a hole or opening in a block or similar device.

reeve the line *v*: to string a wire rope drilling line through the sheaves of the traveling and crown blocks to the hoisting drum.

remotely operated vehicle (ROV) *n*: in offshore operations, an underwater device controlled from a vessel on the water's surface that is used to inspect subsea equipment, such as a pipeline, and that can be used in place of or in conjunction with diving personnel.

reservoir *n*: a subsurface, porous, permeable rock body in which oil and/or gas has accumulated. Most reservoir rocks are limestones, dolomites, sandstones, or a combination. The three basic types of hydrocarbon reservoirs are oil, gas, and condensate. An oil reservoir generally contains three fluids—gas, oil, and water—with oil the dominant product. In the typical oil reservoir, these fluids occur in different phases because of the variance in their densities. Gas, the lightest, occupies the upper part of the reservoir rocks; water, the lower part; and oil, the intermediate section. In addition to its occurrence as a cap or in solution, gas may accumulate independently of the oil; if so, the reservoir is called a gas reservoir. Associated with the gas, in most instances, are salt water and some oil. In a condensate reservoir, the hydrocarbons may exist as a gas, but, when brought to the surface, some of the heavier ones condense to a liquid.

reservoir drive mechanism *n*: the process in which reservoir fluids are caused to flow out of the reservoir rock and into a wellbore by natural energy. Gas drives depend on the fact that, as the reservoir is produced, pressure is reduced, allowing the gas to expand and provide the driving energy. Water-drive reservoirs depend on water pressure to force the hydrocarbons out of the reservoir and into the wellbore.

rig *n*: the derrick or mast, drawworks, and attendant surface equipment of a drilling or workover unit.

rigid platform rig *n*: an offshore platform that does not move with the motion of the wind and sea.

rig manager *n*: an employee of a drilling contractor who is in charge of the entire drilling crew and the drilling rig. Also called a toolpusher, drilling foreman, rig supervisor, or rig superintendent.

rig superintendent *n*: see *toolpusher*.

rig up *v*: to prepare the drilling rig for making hole, i.e., to install tools and machinery before drilling is started.

riser joint *n*: see *riser pipe*.

riser pipe *n*: the pipe and special fittings used on floating offshore drilling rigs to establish a seal between the top of the wellbore, which is on the ocean floor, and the drilling equipment, located above the surface of the water. A riser pipe serves as a guide for the drill stem from the drilling vessel to the wellhead and as a conductor of drilling fluid from the well to the vessel. The riser consists of several sections of pipe and includes special devices to compensate for any movement of the drilling rig caused by waves. Also called marine riser pipe.

roll *n*: the angular motion of a ship or floating offshore drilling rig as its sides move up and down. *v*: to move from side to side.

rotary helper *n*: a worker on a drilling or workover rig, subordinate to the driller, whose primary work station is on the rig floor. On rotary drilling rigs, there are at least two and usually three or more rotary helpers on each crew. Sometimes called floorhand, roughneck, or rig crew member.

rotary system *n*: see *rotating components*.

rotary table *n*: the principal component of a rotary, or rotary machine, used to turn the drill stem and support the drilling assembly. It has a beveled gear arrangement to create the rotational motion and an opening into which bushings are fitted to drive and support the drilling assembly.

rotating components *n pl*: those parts of the drilling or workover rig that are designed to turn or rotate the drill stem and bit—swivel, kelly, kelly bushing, master bushing, and rotary table.

roughneck *n*: see *rotary helper*.

roustabout *n*: 1. a worker on an offshore rig who handles the equipment and supplies that are sent to the rig from the shore base. The head roustabout is very often the crane operator. 2. a worker who assists the foreman in the general work around a producing oilwell, usually on the property of the oil company. 3. a helper on a well servicing unit.

roustabout foreman *n*: head roustabout; supervisor of the roustabout crew.

ROV *abbr*: remotely operated vehicle.

S

salt dome *n*: a dome that is caused by an intrusion of rock salt into overlying sediments. A piercement salt dome is one that has been pushed up so that it penetrates the overlying sediments, leaving them truncated. The formations above the salt plug are usually arched so that they dip in all directions away from the center of the dome, thus frequently forming traps for petroleum accumulations.

satellite well *n*: usually a single well drilled offshore by a mobile offshore drilling unit to produce hydrocarbons from the outer fringes of a reservoir that cannot be produced by primary development wells drilled from a permanent drilling structure (such as a platform rig). Sometimes, several satellite wells will be drilled to exploit marginal reservoirs and avoid the enormous expense of erecting a platform.

sediment and water (S&W) *n*: a material coexisting with, yet foreign to, petroleum liquid and requiring a separate measurement for reasons that include sales accounting. This foreign material includes free water and sediment (dynamic measurement) and/or emulsified or suspended water and sediment (static measurement). The quantity of suspended material present is determined by a centrifuge or laboratory testing of a sample of petroleum liquid.

sedimentary rock *n*: a rock composed of materials that were transported to their present position by wind or water. Sandstone, shale, and limestone are sedimentary rocks.

seismic section *n*: see *record section*.

seismic survey *n*: an exploration method in which strong low-frequency sound waves are generated on the surface or in the water to find subsurface rock structures that may contain hydrocarbons. The sound waves travel through the layers of the earth's crust; however, at formation boundaries some of the waves are reflected back to the surface where sensitive detectors pick them up. Reflections from shallow formations arrive at the surface sooner than reflections from deep formations, and since the reflections are recorded, a record of the depth and configuration of the various formations can be generated. Interpretation of the record can reveal possible hydrocarbon-bearing formations.

self-elevating drilling unit *n*: an offshore drilling rig, usually with a large hull. It has a mat or legs that are lowered to the seafloor and a main deck that is raised above the surface of the water to a distance where it will not be affected by the waves. Also called a jackup drilling rig.

semisubmersible *n*: see *semisubmersible drilling rig*.

semisubmersible drilling rig *n*: a floating offshore drilling unit that has pontoons and columns that, when flooded, cause the unit to submerge in the water to a predetermined depth. Living quarters, storage space, and so forth are assembled on the deck. Semisubmersible rigs are either self-propelled or towed to a drilling site and anchored or dynamically positioned over the site, or both. In shallow water, some semisubmersibles can be ballasted to rest on the seabed. Semisubmersibles are more stable than drill ships and ship-shaped barges and are used extensively to drill wildcat wells in rough waters such as the North Sea. Two types of semisubmersible rigs are the bottle-type semisubmersible and the column-stabilized semisubmersible. See *floating offshore drilling rig*.

separator *n*: a cylindrical or spherical vessel used to isolate the components in streams of mixed fluids. See *oil and gas separator*.

shale shaker *n*: a vibrating screen used to remove cuttings from the circulating fluid in rotary drilling operations. The size of the openings in the screen should be carefully selected to be the smallest size possible that will allow 100 percent flow of the fluid. Also called a shaker.

shaped charge *n*: a relatively small container of high explosive that is loaded into a perforating gun. Upon detonation, the charge releases a small, high-velocity stream of particles (a jet) that penetrates the casing, cement, and formation.

sheave *n*: (pronounced *shiv*) a grooved pulley.

ship-shaped barge *n*: a floating offshore drilling structure that is towed to and from the drilling site. The unit has a streamlined bow and squared-off stern, a drilling derrick usually located near the middle of the barge, and a moon pool below the derrick through which drilling tools pass to the seafloor. Ship-shaped barges are most often used for drilling wells in deep, remote waters. See *floating offshore drilling rig*.

shut in *v*: 1. to close the valves on a well so that it stops producing. 2. to close in a well in which a kick has occurred.

side-pocket mandrel *n*: see *gas-lift mandrel*.

sidewall sampler *n*: used to obtain a core sample. The sampler is activated to fire several small cylinders into the wall of the hole. The cylinders penetrate a short distance into the rock and cut a small core.

single-point buoy mooring system (SPBM) *n*: an offshore system to which the production from several wells located on the seafloor is routed and to which a tanker ship ties up in order to load the produced oil. The tanker is moored to a single point on the buoy and is thus free to rotate around the buoy, depending on wind and current directions.

sinker bar *n*: a heavy weight or bar placed on or near a lightweight wireline tool. The bar provides weight so that the tool will lower properly into the well.

skid off *n*: a jackup drilling rig that can slide its drill floor structure off the jackup barge and onto another structure.

slip joint *n*: see *telescopic joint*.

slips *n pl*: wedge-shaped pieces of metal with serrated inserts (dies) or other gripping elements, such as serrated buttons, that suspend the drill pipe or drill collars in the master bushing of the rotary table when it is necessary to disconnect the drill stem from the kelly or from the top-drive unit's drive shaft. Rotary slips fit around the drill pipe and wedge against the master bushing to support the pipe. Drill collar slips fit around a drill collar and wedge against the master bushing to support the drill collar. Power slips are pneumatically or hydraulically actuated devices that allow the crew to dispense with the manual handling of slips when making a connection.

sloughing (pronounced "sluffing") *n*: see *caving*.

sonde *n*: a logging tool assembly, especially the device in the logging assembly that senses and transmits formation data.

sound generator *n*: in exploration, any device that, when activated by an operator, creates a loud, low frequency sound that penetrates layers of rock.

SPBM *abbr*: single-point buoy mooring.

spread mooring system *n*: a system of rope, chain, or a combination attached to anchors on the ocean floor and to winches on the structure to keep a floating vessel near a fixed location on the sea surface.

spud can *n*: a cylindrical device, usually with a pointed end, that is attached to the bottom of each leg of a jackup drilling unit. The pointed end of the spud penetrates the seafloor and helps stabilize the unit while it is drilling.

spud in *v*: to begin drilling; to start the hole.

Squnch Joint™ *n*: a special threadless tool joint for large-diameter pipe, especially conductor pipe, sometimes used on offshore drilling rigs. When the box is brought down over the pin and weight is applied, a locking device is actuated to seat the joints. Because no rotation is required to make up these joints, their use can save time when the conductor pipe is being run. Squnch Joint is a registered trademark.

stand *n*: the connected joints of pipe racked in the derrick or mast during a trip. The usual stand is about 90 feet (about 27 metres) long, which is three lengths of drill pipe screwed together (a thribble).

station bill *n*: on offshore drilling rigs and production platforms, a poster that gives duties and places for each individual on the rig or platform for various types of emergencies. Every person on the rig or platform should be familiar with the station bill.

steam coil *n*: a pipe, or set of pipes, within an emulsion settling tank through which steam is passed to warm the emulsion and make the oil less viscous. See *fire tube*.

steel-jacket rigid platform rig *n*: a rigid offshore drilling platform used to drill development wells. The foundation of the platform is the jacket, a tall vertical section made of tubular steel members. The jacket, which is usually supported by piles driven into the seabed, extends upward so that the top rises above the waterline. Additional sections that provide space for crew quarters, the drilling rig, and all equipment needed to drill are placed on top of the jacket. See *platform rig*.

stinger *n*: 1. a cylindrical or tubular projection, relatively small in diameter, that extends below a downhole tool and helps guide the tool to a designated spot (such as into the center of a portion of stuck pipe). 2. a device for guiding pipe and lowering it to the water bottom as it is being laid down by a lay barge.

string *n*: the entire length of casing, tubing, sucker rods, or drill pipe run into a hole.

submersible *n*: 1. a two-person submarine used for inspection and testing of offshore pipelines. 2. a submersible drilling rig.

submersible drilling rig *n*: a mobile bottom-supported offshore drilling structure with several compartments that are flooded to cause the structure to submerge and rest on the seafloor. Submersible rigs are designed for use in shallow waters to a maximum of 175 feet (53.4 metres). Submersible drilling rigs include the posted barge submersible, the bottle-type submersible, and the arctic submersible. See *bottom-supported offshore drilling rig.*

subsea blowout preventer *n*: a blowout preventer placed on the seafloor for use by a floating offshore drilling rig.

subsea completion system *n*: where equipment that controls the flow of hydrocarbons from the well is placed on or below the seafloor.

subsea template *n*: a device placed on the seafloor to facilitate the production of wells. When a template is used, the wells are drilled through the template and are completed and produced on it. Since the erection of a platform to produce the wells is not necessary, marginal offshore fields can sometimes be produced because the expense of erecting the platform is avoided.

surface casing string *n*: see *surface pipe.*

surface completion *n*: where equipment that controls the flow of hydrocarbons from the well is placed above the waterline.

surface pipe *n*: the first string of casing (after the conductor pipe) that is set in a well. It varies in length from a few hundred to several thousand feet (metres). Some states require a minimum length to protect freshwater sands. Compare *conductor pipe.*

surfactant *n*: a substance that affects the properties of the surface of a liquid or solid by concentrating on the surface layer. Surfactants are useful in that they ensure that the surface of one substance or object is in thorough contact with the surface of another substance.

swivel *n*: a rotary tool that is hung from the rotary hook and traveling block to suspend and permit free rotation of the drill stem. It also provides a connection for the rotary hose and a passageway for the flow of drilling fluid into the drill stem.

syncline *n*: a downward, trough-shaped configuration of folded, stratified rocks. Compare *anticline.*

T

tanker ship *n*: in certain offshore operations, a ship designed to transport oil from its point of production offshore to a facility on land; tanker ships are often employed when oil cannot be pipelined to shore.

telescopic joint *n*: a device used in the marine riser system of a mobile offshore drilling rig to compensate for the vertical motion of the rig caused by wind, waves, or weather. It consists of an inner barrel attached beneath the rig floor and an outer barrel attached to the riser pipe and is an integrated part of the riser system.

temporary guide base *n*: the initial piece of equipment lowered to the ocean floor once a floating offshore drilling rig has been positioned on location. It serves as an anchor for the guidelines and as a foundation for the permanent guide base and has an opening in the center through which the bit passes. It is also called a drilling template.

tensioner system *n*: a system of devices installed on a floating offshore drilling rig to maintain a constant tension on the riser pipe, despite any vertical motion made by the rig. The guidelines must also be tensioned, and a separate tensioner system is provided for them.

tension-leg platform rig *n*: a compliant offshore drilling platform used to drill development wells. The platform, which resembles a semisubmersible drilling rig, is attached to the seafloor with tensioned steel tubes. The buoyancy of the platform applies tension to the tubes. See *platform rig.*

TFL *abbr*: through-the-flow-line.

through-the-flow-line (TFL) equipment *n*: any equipment designed to be pumped down a completed well to effect a repair, modify the well's flow, or for other reasons.

thruster *n*: see *dynamic positioning.*

tie-in *n*: a collective term for the construction tasks bypassed by regular crews on pipeline construction. Tie-in includes welding road and river crossings, valves, portions of the pipeline left disconnected for hydrostatic testing, and other fabrication assemblies, as well as taping and coating the welds.

tie-in pipeline *n*: a relatively short length of pipeline that connects to a main (trunkline) pipeline.

toolpusher *n*: an employee of a drilling contractor who is in charge of the entire drilling crew and the drilling rig. Also called a drilling foreman, rig manager, rig superintendent, or rig supervisor .

top drive *n*: a device similar to a power swivel that is used in place of the rotary table to turn the drill stem. It also suspends the drill stem in the hole and includes power tongs. Modern top drives combine the elevator, tongs, swivel, and hook.

tour *n*: (pronounced *tower*) a working shift for drilling crew or other oilfield workers. The most common tour is 8 hours; the three daily tours are called daylight, evening (or afternoon), and graveyard (or morning). Sometimes 12-hour tours are used especially on offshore rigs; they are called simply day tour and night tour.

trap *n*: a body of permeable oil-bearing rock surrounded or overlain by an impermeable barrier that prevents oil from escaping.

traveling block *n*: an arrangement of pulleys, or sheaves, through which drilling line is reeved and that moves up and down the derrick or mast. See *block*.

treater *n*: a vessel in which oil is treated for the removal of sediment and water or other objectionable substances with chemicals, heat, electricity, or all three.

trip *n*: the operation of hoisting the drill stem from and returning it to the wellbore. *v*: shortened form of "make a trip". See *make a trip*.

trunkline *n*: a main line.

tubing *n*: small-diameter pipe that is run into a well to serve as a conduit for the passage of oil and gas to the surface.

U

unconformity *n*: 1. lack of continuity in deposition between rock strata in contact with one another, corresponding to a gap in the stratigraphic record. 2. the surface of contact between rock beds in which there is a discontinuity in the ages of rocks.

unloading a well *n*: removing fluid from the tubing in a well, often by means of a swab, to lower the bottomhole pressure in the wellbore at the perforations and induce the well to flow.

V

viscosity *n*: a measure of the resistance of a fluid to flow. Resistance is brought about by the internal friction resulting from the combined effects of cohesion and adhesion. The viscosity of petroleum products is commonly expressed in terms of the time required for a specific volume of the liquid to flow through a capillary tube of a specific size at a given temperature.

W

water drive *n*: the reservoir drive mechanism in which oil is produced by the expansion of the underlying water and rock, which forces the oil into the wellbore. In general, there are two types of water drive: bottom-water drive, in which the oil is totally underlain by water; and edgewater drive, in which only a portion of the oil is in contact with the water.

well completion *n*: the activities and methods necessary to prepare a well for the production of oil and gas; the method by which a flow line for hydrocarbons is established between the reservoir and the surface. The method of well completion used by the operator depends on the individual characteristics of the producing formation or formations. Such techniques include open-hole completions, sand-exclusion completions, tubingless completions, and miniaturized completions.

wellhead connector *n*: a remote-controlled hydraulic clamp located in the main section of the BOP stack.

well logging *n*: the recording of information about subsurface geological formations, including records kept by the driller and records of mud and cutting analyses, core analysis, drill stem tests, and electric, acoustic, and radioactivity procedures.

well service and workover contractor *n*: a company that specializes in offshore well repair.

well servicing *n*: the maintenance work performed on an oil or gas well to improve or maintain the production from a formation already producing. Usually it involves repairs to the pump, rods, gas-lift valves, tubing, packers, and so forth.

wet gas *n*: 1. a gas containing water, or a gas that has not been dehydrated. 2. a rich gas.

wet glycol *n*: glycol that has absorbed water.

widowmaker (slang) *n*: on offshore rigs, a narrow walkway placed between the barge and the platform which allows the crew to move back and forth between the two.

wildcat *n*: a well drilled in an area where no oil or gas production exists. With present-day exploration methods and equipment, about one wildcat out of every seven proves to be productive, although not necessarily profitable. *v*: to drill wildcat wells.

wireline *n*: a small-diameter metal line used in wireline operations; also called slick line.

wire rope *n*: a cable composed of steel wires twisted around a central core of fiber or steel wire to create a rope of great strength and considerable flexibility. Wire rope is used as drilling line (in rotary and cable-tool rigs), coring line, servicing line, winch line, and so on. It is often called cable or wireline; however, wireline is a single, slender metal rod, usually very flexible. Compare *wireline*.